ARMAGEDDON
SCIENCE

ALSO BY BRIAN CLEGG

ARMAGEDDON
SCIENCE

THE SCIENCE OF MASS DESTRUCTION

BRIAN CLEGG

St. Martin's Press   *New York*

www.stmartins.com

Book design by Rich Arnold

Library of Congress Cataloging-in-Publication Data

Clegg, Brian.
    Armageddon science : the science of mass destruction / Brian Clegg.—1st ed.
                p. cm.
    Includes bibliographical references.
    ISBN 978-0-312-59894-5
    1. Science—Social aspects.   2. Research—Moral and ethical aspects.   3. Armageddon.   I. Title.
    Q175.5.C54 2010
    303.48'3—dc22

                                                            2010032524

First Edition: November 2010

10   9   8   7   6   5   4   3   2   1

FOR GILLIAN, REBECCA, AND CHELSEA

CONTENTS

ACKNOWLEDGMENTS

As always, this book would not have been possible without the help and support of my editor, Michael Homler, and my agent, Peter Cox.

I usually list those who have helped me, but they have been numerous, and I would rather just say a big thank-you for all the input I have received.

CHAPTER ONE
MAD SCIENTISTS

||||||||||||||||||||||||||||||||||||||||||||||

> *It was the secrets of heaven and earth that I desired to learn.*
> —Mary Shelley (1797–1851), *Frankenstein* (1818)

Mass destruction—killing on a vast scale—is a uniquely human concern. It's not that other animal species aren't threatened by it. Many have been driven to extinction, and many more now teeter on the brink. But unlike human beings, even the most intelligent animals don't worry about the possibility of being wiped out in a terrible catastrophe. It is only thanks to the human ability to contemplate the future that fears of mass destruction have arisen. As the continued popularity of disaster movies at the box office demonstrates, we are all too aware how, as a race, we might be wiped out.

Mass destruction has, historically, been a natural phenomenon. The Earth has witnessed widespread devastation numerous times, most famously in the destruction of the dinosaurs 65 million years ago. We could still see a similar act of mass destruction in the future that does not require a human hand behind it. But with the introduction of the weapon of mass destruction, the notion is most commonly associated with the work of the mad—or at best, amoral—scientist.

The term "weapons of mass destruction" first appeared in a Christmas sermon by the archbishop of Canterbury in 1937. He encouraged his audience to promote peace. "Who can think without dismay of the fears, jealousies and suspicions which have compelled nations, our own among them, to pile up their armaments," he said. "Who can think without horror of what another widespread war would mean, waged as it would be with all the new weapons of mass destruction."

The archbishop was concentrating on the political will to use such weapons. His was a generation that had lived through the First World War, expecting it to be the "war to end all wars," yet was seeing the rapid buildup of military might in Europe as the Second World War loomed. However responsible politics was for the warfare, though, it goes without saying that scientists would be the ones who made such weapons exist.

It's a truth that can't be avoided. Science itself—or at least, the application of science—has a dark side. Scientists present us with dangerous gifts.

This isn't a new idea, though for a brief period—from Victorian times through to the mid-twentieth century—scientists were seen in quite a different light. New technologies and scientific developments transformed the unpleasant life suffered by the vast majority of the population into a new kind of existence. It was no longer necessary to spend every moment scratching out a living. For the first time, it wasn't just the rich and powerful who had time for leisure and enjoyment of life. Scientists were briefly considered saviors of our race.

These men (and back then they almost all were men) were bold bringers of wonderful new things, Santa Claus and the Easter Bunny rolled into one real package that delivered all year round. All the marvels of electricity, of modern medicine, of new modes of transport and labor-saving devices, were their gift. And we still see echoes of this in TV ads for beauty products, where the person in the white coat is the bringer of magic ingredients that are guaranteed to make you look better and younger.

But the warning of Pandora's box, the dangers inherent in bringing knowledge into the world, could not be held off for long. If you live in a physically dangerous environment, trying new things, finding things out, is a high-risk strategy. If a cave person decided to experiment with a new approach to saber-toothed tigers, patting them on the head instead of sticking them with a spear, she would soon be a one-armed cave person. For most of history, the scientist and his predecessor, the natural philosopher, have been characters of suspicion, closely allied with magicians, sorcerers, and other dabblers in arcane arts. This was not a stereotype that even the wonders of nineteenth- and twentieth-century technology could hold off for long.

Scientists as dangers to the world would return in pulp fiction and cheap movies, where they are often portrayed as barely human. At best, these driven souls are over-idealistic and unworldly. They are what my grandmother would have called "all cleverness and no common sense." They are innocents who don't know—or don't care—what the outcomes of their acts will be. At the nasty end of the spectrum, they are even worse, evil beings filled with a frenzied determination to achieve world domination or to pursue what they see as scientific truth at any cost.

Such two-dimensional, caricature scientists don't care whom they trample to reach their goal. They have a casual disregard for the impact of what they do on human life—or even on the planet as a whole. They are scientific Nazis for whom the end always justifies the means. They are nothing short of monsters in human form.

Practically all the scientists I have ever met are not like this. They are warm, normal people. They have the same concerns as everyone else about the world their children will inhabit, the same worries that preoccupy us all. Admittedly some are unashamed geeks—and if you consider a "geek" anyone who has a sense of wonder about the universe he lives in, it's a group in which I happily proclaim my membership—but they aren't inhuman thinking machines. So where did this idea come from?

Inevitably science fiction has to bear a fair amount of the blame for this portrayal. When the teenage Mary Godwin (soon to become Mary Shelley) first penned *Frankenstein* on a traumatic vacation in an Italian villa, she certainly had in mind that her character was playing God. He admits as much in his confession that opens this chapter, "It was the secrets of heaven and earth that I desired to learn." There is no modesty here—Victor Frankenstein wants to be a master of the universe, and his talkative creation, very different from the shambling, incoherent creature of the movies, spends great swaths of text agonizing over the dangers of this philosophy.

Yet Mary Shelley's Baron Frankenstein is not quite yet the archetypal mad scientist—an expression that would become so common as to be a cliché. It first took those doyens of nineteenth- and twentieth-century science fiction, Jules Verne and H. G. Wells, to show us just how driven their imagined scientists could be. So Verne's Captain Nemo could be a merciless killer, and Wells would give us characters like the Invisible Man, driven insane by his search for knowledge, and Dr. Moreau, who despoiled animal and human alike with his merciless vivisection.

The final nail in the coffin would come with the contribution of Hollywood. Here Victor Frankenstein would be transformed from a thoughtful (if megalomaniac) philosopher to a crazed, wide-eyed freak. On celluloid would be born the evil genius Rotwang in Fritz Lang's stunning silent movie *Metropolis*, and Peter Sellers's darkly humorous portrayal of the appalling Dr. Strangelove. These movie madmen would be joined by the living incarnations of evil comic-book scientific geniuses, from Lex Luthor to the Green Goblin.

Even when a scientist managed to be one of the good guys, such as Doc Brown in the *Back to the Future* series of movies, he still sported the wild hair (Einstein can probably be blamed for this) and semi-irrational behavior of his more dangerous equivalents. Perhaps most telling of all, Hollywood would give us *Forbidden Planet*, with Doctor Morbius and his "monsters from the id"— destructive forces released by expanding the capacity of the human mind.

Forbidden Planet gives us the real clue to the long-term origins of the mad scientist—because *Forbidden Planet* was based on Shakespeare's play *The Tempest*. Dr. Morbius stands in for Prospero, a philosopher whose world is distorted by the forces he has brought into play and which he can no longer control. This was an image with a long history in literary and folk tradition. Often in the early days it was considered that anyone who worked in natural philosophy—what we would call science—was dabbling with magic and dealing with the devil. Such early "mad scientists" were often thought to have their own equivalent of Prospero's magic servant Ariel in the form of talking metal heads.

The earliest Western owner in legend of such a marvelous but terrifying engine was the French cleric Gerbert of Aurillac. By the time this scholarly abbot became Pope Sylvester II in 999, he had already gained the reputation of being a black magician, a fate that seemed to attach itself to anyone with scientific leanings. Gerbert was associated with a talking brass head in one of the anecdotes used by the monk William of Malmesbury to liven up his histories.

He tells us that Gerbert imbued the head with its magical powers by casting it using "a certain inspection of the stars when the planets were about to begin their courses." It was inevitable that Gerbert should have been seen as a necromancer. He was a man who had no qualms about investigating the mysterious workings of nature. Not only did he write books, he also described marvelous devices and drew strange diagrams that looked like magical symbols if you didn't understand the science he was attempting to portray.

Anyone who delved into the workings of nature was regarded with suspicion at a time when the study of natural philosophy was only just beginning to creep back into Europe. Western scientific knowledge was based largely on the work of the Greek philosophers, who from around 500 BC had begun to speculate on the nature of the universe.

With the fall of Greek civilization, what knowledge there had been was lost, or at least mislaid. Hundreds of years later, the Arabs began to discover the remnants of Greek libraries, and set about a painstaking reconstruction of what had been known, to which they added their own insights; but it was only when Christian scholars brought this Greek-Arab amalgam into Latin-speaking Europe that the study of nature began to find acceptance, and then only among the cognoscenti. For the common people it was all still a mystery—with its adepts branded as sorcerers.

By the thirteenth century, the legend of the head had transferred itself from Gerbert to another figure. It was now said to be the creation of the Bavarian Albertus Magnus. Albertus was a great encyclopedist of science, but in England it was Roger Bacon's name that stuck in the popular mind, and by Tudor times the head of brass had moved again, now firmly established as Friar Bacon's property. Bacon too was an early scientist, writing a huge book on the scientific knowledge of the time. He was explicit in his denial of the existence of magic, explaining everything by nature and art (human work), but again his natural philosophy made him considered a magician. After his death, Bacon became the legendary owner of the talking brass head.

Even though the tale of the head is clearly fictional, such was the strength of the legend that it came to influence reality. Two colleges at the venerable Oxford University have in the past claimed Bacon as an alumnus—Merton and Brasenose. Both claims are highly doubtful. Although there was probably some overlap between the existence of Merton College, founded in 1264, and Bacon's second stay in Oxford, Merton was originally dedicated to educating the sons of Walter of Merton's seven very productive sisters. That hardly made it a likely home for Bacon.

It's at Brasenose, though, that the head appears to emerge into real life. The Brasenose story has even less going for it than Merton's, though the college has to be admired for the sheer cheek of its attempt to claim Roger for its own. Brasenose didn't open until 1509, more than two hundred years after Bacon's death. But by then the brass-head legend would have been well established, and it could well be that the "Brasenose" of the name, an oversized brass nose placed over the gatehouse of the college, was believed to be a remnant from the explosion that was said to have destroyed the famous brass head.

By the sixteenth century, the tale of the head and other stories had solidified into the lively, earthy collection of tales called *The Famous Historie of Fryer Bacon*. Soon afterward, Bacon's adventures were brought to life for the public in a play based on the *Historie* entitled *The Honourable History of Friar Bacon and Friar Bungay*, by Robert Greene, a largely forgotten contemporary of William Shakespeare. Greene's Bacon was a Faustian figure—in fact, for a long time *The Honourable History* was thought to be a rip-off of Christopher Marlowe's play *The Tragical History of Dr. Faustus*. Marlowe himself was not above a little plagiarism, using as his source the legend that had built up around the fifteenth-century German scholar and magician Johann Faust.

The real Faust was born in Württemberg around 1480. He went to university, but found the easy life of fortune-telling and conjuring more attractive than his studies and began to travel from town to town, making money wherever and however he could. To boost his reputation he openly boasted that he had sold his soul to the devil, a claim that Martin Luther took seriously enough to brand him a master of demonic powers. Others thought Faust an opportunistic charlatan, but nonetheless a dangerous man to have around. He was thrown out of the city of Ingolstadt in 1528. The municipal records note that "a certain man who called himself Dr. Johann Faust of Heidelberg was told to spend his penny elsewhere, and he pledged himself not to take vengeance on or to make fools of the authorities for this order."

After Faust's death, his reputation started a legend that spread throughout Europe. Marlowe's *Dr. Faustus* tells of a man obsessed with power. In exchange for his soul he gains knowledge and influence. By the end of the story he repents of his actions, even though he is too late to be saved from damnation. It's easy to see an early form of the mad scientist here. Faust didn't care about the thing that really mattered—his soul; instead, the rampant search for knowledge drove him to the devil.

In the whole picture of human beings "playing with forces we don't understand" that dominates these medieval tales, there's an element of reality. Science really does put human existence at risk—not because there are madmen in charge of laboratories, or because knowledge has somehow warped the scientists' brains, but because human beings have a relentless, unstoppable urge to venture into the unknown, an urge that outside of science is generally seen in a much more favorable light.

It's the same spirit that sent the pioneers out to the American West, that *Star Trek* urge to "boldly go where no one has gone before." Inevitably, such exploration can take us into danger. We do our best to keep that danger to a minimum, but we can't make the risk entirely go away. Science will always involve an element of danger, just as being human always involves an element of danger. And as our science gets deeper, more fundamental, then the potential scale of that danger grows.

This is why in 2008 a group tried to get a court injunction to prevent the turning on of the biggest machine human beings have ever contemplated building. The group was convinced that throwing that switch would do nothing less than destroy the world. It would, they believed, not just kill off the human race, but threaten the whole existence of reality as we know it.

CHAPTER TWO
BIG BANGS AND BLACK HOLES

||||||||||||||||||||||||||||||||||||||||||||||||||||||||||||||||||||||||||||||

> *It is becoming clear that in a sense the cosmos provides the only*
> *laboratory where sufficiently extreme conditions are ever achieved*
> *to test new ideas on particle physics. The energies in the Big Bang*
> *were far higher than anything we can ever achieve on Earth.*
>
> —Martin Rees (1942–), in *A Passion for Science*,
> ed. Lewis Wolpert and Alison Richards (1988)

When radio astronomer Martin Rees made the comments at the top of the page in 1988, suggesting that we would never be able to produce dramatic enough conditions on the Earth to reach the extremes required for some experiments in subatomic physics, it was a fair comment. But over twenty years later, the capabilities of our particle colliders have greatly advanced beyond anything possible back then. We can't create cosmic chaos on the scale of the big bang—which is probably just as well—but locally the conditions will be approaching the extremes of creation. And that's a worry. Because there's something about experimental particle physics that brings out the mad scientist in anyone.

> **Big bang warning**—The big bang is one of several theories
> of how the universe came into being, and although the

> current evidence mostly supports the big bang theory
> (after that theory has been severely tweaked to fit the
> data), that evidence is very indirect, and there are other
> theories that also fit the data just as well. I want to make
> this point because to keep things simple, I will be refer-
> ring to the big bang as if it definitely happened, but there
> does remain a considerable degree of uncertainty about
> the concept.

It's not really surprising that particle physics and the image of the mad scientist go well together. We are combining the most childish form of science with far and away the most expensive toys around—it's hardly a shock that the effect can be terrifying. Why childish? Because the way children often get a handle on reality is by hitting things and seeing what happens. What can be more childish than attempting to discover how something works by breaking it? To be clear about this, let's re-create the thought experiment of eighteenth-century philosopher William Paley.

Paley imagined discovering a pocket watch when out on a walk on the English heath. Suppose, he said, you had never seen a timepiece before. By carefully examining the watch, by exploring the complexity of its manufacture and guessing at its function and how it was constructed, it would be reasonable to deduce that this wasn't a natural phenomenon like a rock, or even a heather plant. You would realize that this was surely a designed object. And for something to be designed, it requires a designer. We can use inductive reasoning to say that the watch implies the existence of a designer.

For Paley, this was an analogy that could be extended to life on Earth, which in his mind also required a designer. He commented:

> Every indication of contrivance, every manifestation of design, which existed in the watch, exists in the works of nature; with the difference, on the side of nature, of being greater or more, and that in a degree which exceeds all computation.

The mechanisms behind modern evolutionary theory, including natural selection, have proved Paley wrong in drawing conclusions about life from the model of the watch on the heath. But the approach generally taken by science would still be to carefully analyze how the watch works so that we can determine just what it is and how it functions. This is not the approach taken by particle physics. To make this a particle-physics analogy, scientists would take a sledgehammer and smash the watch as hard as they could, taking high-speed photographs as they did so to capture the trajectories of the gears, springs, and other components that flew out of it. From these photographs they would attempt to work out just what had been going on in the watch.

Modern particle physics is all about finding bigger and better ways to smash particles together. There is no careful attempt to analyze the nature of the particles—this is not like the work of, say, an archaeologist, painstakingly brushing away fragments of dirt. If a particle physicist were an archaeologist she would excavate a site with dynamite. These scientists accelerate particles mercilessly until they are traveling at nearly the speed of light, then slam them into one another in head-on collisions. It's like the ultimate dream of every small child who has smashed one toy car into another, combined with a teenage enthusiasm for vast machines and underground laboratories—a particle accelerator would make an ideal James Bond movie set.

To see the potential for mad science at its most deadly, we have to travel to an out-of-the-way country location near Geneva, Switzerland. It's there that the Conseil Européen pour la Recherche Nucléaire (CERN) is located. CERN is a vast international research organization that has built the biggest machine ever envisaged by human beings: the Large Hadron Collider (LHC). "Large" is a totally inadequate adjective. The LHC is immense. Not that it's obvious to the passerby. In best Bond villain style, the scientists behind this vast mechanism have constructed it deep underground.

Imagine a 27 kilometer (17 mile) long circular tunnel, easily large enough to drive a car through at over 3.8 meters (twelve feet) wide. Through the center of the tunnel runs an immense metal tube, straddling the border between Switzerland and France. The travelers that journey through this tube would baffle immigration officials, switching from country to country thousands of times a second. This is a nightmare carousel where charged particles are given repeated pushes by vast electromagnets the size of houses. This machine requires the kind of power supply that is needed to run an entire city. Time after time, the particles fly around the circuit, their route carefully controlled by computer until they are brought into head-on collisions in one of the building-sized particle detectors.

CERN had already become well known before the LHC was

planned to go online in 2008. One of CERN's staff, the British computer scientist Sir Tim Berners-Lee, had thought of a different way to use the then new Internet to share information among geographically remote laboratories. He called his invention—rather grandiosely, since it was accessible at only a few sites to begin with—the World Wide Web. But how right this anything-but-mad scientist was with the tongue-in-cheek name he gave to his invention.

And then there's another, more recent contribution to CERN's fame, which has come from a surprising source: the novelist Dan Brown. One of Brown's novels, *Angels and Demons* (made into a movie in 2009), is partly set in CERN. At the heart of the novel is a possible source of terrible destruction, antimatter. Although many of us first met antimatter as a power source on *Star Trek*, it's a real enough concept. Antimatter is the same as ordinary matter, but the particles that make it up have the opposite electrical charge of those in conventional matter.

Where, for example, an electron has a negative charge, the antimatter equivalent, the antielectron (usually called a positron), has a positive charge. There are similar, antimatter equivalents of all the particles. When two opposite-charged antimatter particles—an electron and a positron, for example—are brought together, they are attracted, smash into each other, and are destroyed.

The particles' mass is converted into energy, and though particles, like electrons, are very light, Einstein's famous equation $E = mc^2$ tells us that the energy produced will be equal to the mass of the particles multiplied by the square of the speed of light. That's a big number. A pound of antimatter wiping out a pound of normal matter would produce about as much energy as a typical power station pumps out during six years of running.

This kind of explosive interaction doesn't happen when a negative electron orbits a positive proton in an atom, because there are nuclear forces in place to prevent annihilation, but no such force protects matter and antimatter. We don't generally see antimatter on the Earth, because it would disappear in an instant, taking out an equivalent amount of matter and producing a dramatic explosion. But it can be manufactured in the laboratory—and it has been at CERN.

In Dan Brown's *Angels and Demons,* antimatter produced at CERN is used to make a devastating bomb, which fanatics plan to use to blow up the Vatican. We are told in the novel that just one gram of antimatter, little more than a pinch of the material, will produce an explosion equivalent to the twenty-kiloton atomic bomb that devastated Hiroshima at the end of the Second World War.

If anything, this underestimates the devastating power of antimatter. It would take less than half a gram to have that effect. Remember, when antimatter collides with normal matter, the mass of every atom is converted into energy according to $E = mc^2$, and that's a lot of energy. To make the concept of antimatter as a weapon even more remarkable, it has been claimed that an antimatter bomb is a clean bomb that destroys without producing the radioactive devastation that accompanies a nuclear weapon, particularly a hydrogen bomb. This seems to have been enough to get some members of the U.S. Air Force excited, and in the early years of the twenty-first century, rumors started to spread that the Air Force was building an antimatter weapon.

On March 24, 2004, attendees at an otherwise conventionally tedious conference were suddenly jerked upright in their seats by what they heard. Speaking at the NASA Institute for Advanced Concepts conference in Arlington, Virginia, Kenneth Edwards, director of the Air Force's revolutionary munitions team, was about to explain just how dangerous antimatter was.

He told his audience of the potentially devastating power of this substance—even if only tiny amounts were present. As a graphic example, he considered the 1995 Oklahoma City bombing that had left 168 dead. To produce the same devastation, he said, would take just 50 *millionths* of a gram of antimatter in the form of positrons. Like Timothy McVeigh's bomb, it would produce an equivalent blast to over two tons of TNT.

There was an immediate press uproar. Four months later, Edwards's team was still pushing antimatter, saying everyone was "very excited about the technology"—and then came silence that has continued to the present day. Were devastating weapons that could destroy silently with a tiny power source about to be unleashed? Was this an *X-Files*–style conspiracy in which everyone who knew about this potentially devastating weapon had been silenced? It's unlikely. Instead, what had happened was the sober realization that the antimatter bomb was a pipe dream. A fantasy.

Dan Brown's book *Angels and Demons* may well have had something to do with the sudden upsurge of interest in antimatter as weaponry. Interest in the story, which features the same protagonist as Brown's sequel, *The Da Vinci Code,* surged with that book's massive success. Something similar happened in 2009 with the release of the *Angels and Demons* movie. And that's unfortunate, because though a lot of the plot is typical puzzle- and action-driven hokum, there is a fair amount of "science" in the book that is simply wrong.

We can allow Brown some carelessness with the truth in the novel—it is fiction, after all—but up front in the book is a section labeled "FACT"; sadly, even this is well adrift of the truth. CERN, we are told, "recently succeeded in producing the first particles of antimatter." Well, no, that happened way back in 1932 when American scientist Carl D. Anderson discovered positrons—antimatter electrons. What Brown might have meant was that CERN has recently (in 1995, five years before *Angels and Demons* was published) created antihydrogen, antimatter *atoms,* as opposed to the more controllable charged antimatter particles like positrons.

Antimatter, we are told by Brown, is "the most powerful energy source known to man." That's okay as far as it goes. "A single gram of antimatter contains the energy of a 20-kiloton nuclear bomb—the size of the bomb they dropped on Hiroshima." Half right—this is the energy you'd get if you converted one gram of antimatter into pure energy. But that's not how it works. You have to combine that gram of antimatter with a gram of matter, getting twice the amount of energy out. Is it coincidental that Kenneth Edwards made the same mistake four years later? Perhaps not.

However, there is a big assumption being made, even if you could come up with a gram of antimatter by waving a magic wand. Explosions aren't caused solely by the amount of energy in something, but also by how fast it is released. The difference between something burning in a controlled way and something exploding is just a matter of timing. When a quantity of a substance all burns at once, we call it an explosion. But having the built-in energy doesn't guarantee that the substance will explode.

Take a simple example: which is more explosive, TNT or gasoline? The TNT, of course. That's why we use it to blow things up.

But which has more chemical energy locked away in it? The gas. Weight for weight, gasoline delivers fifteen times as much energy as TNT. It's just that the TNT burns a heck of a lot faster. As we've never combined a gram of antimatter with a gram of matter, we don't know if it would fizzle away a particle at a time, or go up in a moment, producing that time-compacted conversion to energy that is an explosion.

Back with the "facts" in *Angels and Demons,* we're told that (exploding) antimatter produces no pollution and no radiation, making it a clean source of energy. This is just plain wrong. Incredibly, dramatically wrong. When antimatter combines with matter it pumps out gamma rays. These are ultrahigh-energy electromagnetic radiation, far more powerful than X-rays, and will do devastating damage to living tissue. Gamma rays produce most of the lasting damage that arises from nuclear fallout.

Finally, and correctly, in his "FACT" section, Brown says, "Until recently antimatter has only been created in very small amounts," but he goes on to say that CERN now has its new Antiproton Decelerator, "an advanced antimatter production facility that promises to create antimatter in much larger quantities."

The Antiproton Decelerator does exist—it's a mechanism to slow down antiprotons to make them controllable so they don't slam into matter and wink out of existence. But it doesn't actually create those antiprotons. More significantly, although CERN can now make considerably more antiprotons than were first produced at the site, we're still only talking about around a million particles at a time. You would need around 100 trillion times that to come close to 50 millionths of a gram, the amount Kenneth Edwards put forward as being equivalent to the Oklahoma bomb.

There simply isn't a means on Earth of producing antimatter anywhere fast enough to meet the needs of an antimatter weapon program. At current rates it would take millions of years to make a single gram of antimatter. This is a problem Dan Brown recognizes in his book. He likens it to building an oil rig to produce a single barrel of oil. This would, of course, make for very slow production of oil that was very expensive.

Unfortunately, Brown then suggest that, just like with the oil rig, all we need to do is produce lots more antimatter from the same technology, because once we've got over the construction cost, it's relatively cheap to produce. This is true of oil, but not of antimatter. It still takes much too much money to make antimatter for it to be an energy source as Brown suggests it could be. But there's a bigger problem—efficiency.

It's easy to think of antimatter as a very efficient source of power, because it produces vastly more energy per gram than any other source, including conventional nuclear energy. In that sense, it is efficient. But there's another meaning of efficiency where it fails woefully. For an energy source to be practical, it has to produce more energy than you put in to produce the source in the first place.

To make antimatter takes considerably more energy than is released when the antimatter annihilates. Imagine it took two barrels of oil's worth of energy to produce one barrel of the black gold. You wouldn't bother to make the oil—it wouldn't be commercially viable. Similarly, you would never use antimatter as an energy source. Unless you've already got some antimatter, it isn't an energy source, it's an energy sink. You would be better off directly using the energy required to make it than you would using antimatter.

And then you would need to put the antimatter somewhere. This is anything but trivial—in fact, it's a huge challenge. You need to contain a substance that destroys matter on contact. It's not like storing a strong acid, where you just have to find the right kind of resistive material. It doesn't matter what kind of container you use, it will be zapped by the antimatter. So how do scientists cope with the tiny quantities they currently work with? They use insubstantial containers—vessels that make use of electromagnetic repulsion to hold antimatter in place.

Visitors to Shanghai in China will have experienced something similar. They can ride on a maglev train from Pudong airport to the city. This train has no wheels. It is lifted off the rails by magnetic levitation. Just as two toy magnets will push away from each other, strong electromagnets in the train and the track repel one other, holding the train cars floating a tiny distance above the track.

Containers (also called traps) for antimatter work in a similar way. The air is removed from the container, leaving as good a vacuum as possible. Strong electromagnetic fields are set up, acting from different directions, to produce a pinch point in the center of the container where anything with the same charge would be forced to sit by repulsion. Charged antimatter particles—typically positrons (antielectrons) or antiprotons—are injected into the trap and sit suspended by the field, not touching any normal matter.

This is described quite well in *Angels and Demons* (though the traps employed are too pretty and cinematic in their appearance). But Brown then blows the whole idea when he has an antimatter bomb hidden in the Vatican. When the Swiss Guards want to search for the device, they are told by the book's heroine that the antimatter has the chemical signature of pure hydrogen.

Now, it is entirely possible to make antihydrogen, with an anti-proton as its nucleus and a positron in place of the normal electron. As we've already seen, such antiatoms have been made at CERN since 1995. But an antiatom will inevitably be destroyed within a tiny fraction of a second of its creation, because there is no way of storing it. Unlike charged particles, the neutral antiatom can't be held in an electromagnetic field trap. There is no way to get a grip on it. So the antiatoms annihilate almost immediately. If, as Brown suggests, the antimatter in the bomb had the chemical signature of hydrogen, it couldn't be held in safe suspension. It just wouldn't work.

Even if the antimatter bombers had been more sensible and used positrons or antiprotons, they would have had a problem. Imagine what happens as we pour more and more positrons into an electromagnetic trap. Each of those positrons has a positive charge. They will all be fighting to get away from one another. The more we put in, the harder it is to keep them in place. Only tiny amounts of antimatter—perhaps a few million particles—can be stored in a trap before the repulsion becomes too great and they start to leak out.

There wouldn't be such a problem with antiatoms, like the anti-hydrogen in *Angels and Demons,* because the atoms aren't charged. Large amounts could be squeezed into an antimatter bottle, just as much as normal hydrogen could be stored in a bottle made of ordinary matter. But there is no physical or magnetic container on the Earth that could keep those antiatoms in place and stop them from annihilating immediately with the matter around them. We can't make an antimatter bottle.

We do have an example of handling a significantly bigger

amount of dangerous charged particles, though—in a fusion toko-mak. This is a vast magnetic container, shaped like a ring dough-nut, that is used to contain the sunlike plasma that it is hoped will one day be at the heart of fusion power plants. Although the plasma isn't antimatter, it would be very destructive if it came into contact with the walls of the tokomak. And there is considerably more matter in a tokomak like the Joint European Torus at Culham, En-gland, than in any antimatter trap. But a tokomak is a big structure, the size of a large office building. It isn't exactly portable, and cer-tainly couldn't be used to transport an antimatter bomb around.

Before antimatter can be stored, however small the quantity, it needs to be produced. At any one time there is a small amount of antimatter on the Earth from natural sources—both from emis-sions from nuclear reactions and from the impact of cosmic rays on our atmosphere. Usually, such particles are destroyed so quickly that we can't do anything with them; but catch them quickly enough and they are valuable in a medical device, the positron-emission tomography or PET scanner.

For the PET scanner to work, a chemical based on a large mol-ecule that the human body will process—typically the sugar fluo-rodeoxyglucose—is injected into the bloodstream. This carries with it small quantities of a radioactive isotope with a short half-life like carbon 11 or fluorine 18, which emit positrons as the nu-cleus decays. The large molecule is carried by the body into the tissue, taking the tracer isotope with it.

As the radioactive substance emits positrons, these antimatter electrons immediately interact with their normal-matter equiva-lents and are converted to energy in the form of a pair of high-energy gamma ray photons. These shoot off in opposite directions until they reach the doughnutlike detector that is around the relevant section of the patient's body. Here, the gamma rays interact with a substance called a scintillator, which is stimulated by the gamma ray photons into giving off a burst of lower-energy light.

This is the same approach that was taken when nuclear decay was first discovered—but then the scintillator had to be observed in a darkened room by eye or through a microscope. In the PET scanner, that ring also contains electronics designed to take a small amount of light and amplify it, converting the tiny flashes that appear simultaneously on both sides of the ring into a signal that can be registered on a computer and built into a "slice-by-slice" image of the cross section of the part of the body of interest.

The PET scanner is an example of using a seminatural source of antimatter. The antimatter is produced in a natural fashion from the breakdown of the atomic nucleus, but those unstable, short-lived isotopes are produced artificially using a device such as a cyclotron, which is a small particle accelerator (typically the size of an SUV) located at or near the hospital where the scanner is to be used.

A more dramatic possibility for a fully natural source of antimatter is that there could be a whole universe of it out there, if only we could get access to it. When matter initially formed after the big bang, there was no particular reason why it should be purely conventional matter. In the incredibly high-energy state immediately after the big bang, energy would constantly be converted into pairs of matter and antimatter particles. In principle, there should have been equal amounts of matter and antimatter, which then would eventually wipe each other out, leaving a universe full of energy alone.

That this didn't happen is usually explained by assuming that very subtle differences in the properties of matter and antimatter meant that there was a tiny extra percentage of matter—everything else was then wiped out, leaving only this excess. This theory, devised by Andrey Sakharov, the Russian physicist better known for being a political dissident, suggests that as a little as one particle in a billion survived the vast matter/antimatter wipeout. But that was enough.

Some have speculated, though, that instead of the antimatter being destroyed in those early days of existence, the universe in some way became segmented, and that there are vast pockets of antimatter out there—perhaps on the same scale as our own observable universe. If the two ever came into contact, the result would be an outpouring of energy that would make every supernova ever seen combined look like a match being struck.

On a more practical level, antimatter is usually made in the laboratory as the product of a high-energy collision, for instance by shooting protons at a metal target. The antimatter doesn't come from the matter but rather from the energy of collision. Just as happened with the seething energy after the big bang, a large amount of energy can spontaneously convert into a pair of particles—one matter, one antimatter.

It's Einstein's $E = mc^2$ equation working in reverse. Here energy is being converted into matter. The newly created pair of particles tend to fly off wildly, and normally the antimatter half would very soon smash into a matter particle and disappear back into energy. When antimatter is produced in the lab, though, the antimatter particle is braked by sending it through a sea of charged particles, which absorb energy, slowing it down, before it can be captured in a magnetic field and stored. This is a delicate process—too much damping and the antiparticle will be annihilated in the braking medium, but get it just right and you've caught yourself an antiparticle.

Sadly for Dan Brown fans, if not for the survival of the world, the *Angels and Demons* scenario fails on practically every level. We can't make enough antimatter, we can't keep more than a tiny amount of charged antimatter in a trap, any store that could hold a usable amount would not be transportable, and we can't keep uncharged antimatter at all. The antimatter bomb, or any other form of antimatter weapon, is not likely to emerge from CERN, nor to be a danger to anyone in the foreseeable future.

But the production of antimatter is a relatively mundane and small-scale aspect of CERN's potential as a source of destructive power. Now that the Large Hadron Collider has been brought to its full capacity, it is generating energies with a concentration that has never before been produced by human beings. There is no danger that this will destroy the whole universe and start things over again in a repeat of the big bang—although the energy is remarkably high, it is infinitesimal in scale compared to the real big bang, and there have been plenty of natural cosmic events since the early days of the universe with much more energy—but there are two possibilities for the LHC to produce destructive materials that could result in devastation.

All the scientists involved at CERN are very clear that these apparent sources of danger are nothing to worry about. They play down the risks. But then, they would. To some outside observers there seems plenty of reason for fear. As we will see later, an attempt has been made to take out an injunction against CERN to prevent the scientists there from destroying the universe.

The most likely dangerous-sounding products of the Large Hadron Collider are tiny, particle-sized black holes. Let's take a step back from the Hollywood image of a black hole for a moment and understand what one is before we start to worry about what it could do. If you believed Hollywood, the black hole would be like an unstoppable space vacuum cleaner, sucking in anything and everything in its path, capable of destroying an entire galaxy as it sucks more and more material into its inescapable gravity field, all the time increasing its gravitational strength.

Before looking at what the LHC could create, it's worth noting that black holes are theoretical constructs. No one has seen one directly. They've certainly not been experimented on. All we have is theory and indirect observation. It's strong theory, and it's very probably true, but there are alternatives that would explain the phenomena we believe are caused by black holes without the real things existing. To be fair to black-hole supporters, though—and that's the vast majority of astronomers and cosmologists—there is a better basis for black holes' existence than there is for many other cosmological phenomena.

The reasoning behind our assumption that black holes exist is twofold. They seem to be an inevitable conclusion of certain physical processes, even though those processes don't have to have ever happened in reality; and various observations from deep space seem to suggest that black holes are more than just speculation.

The physical processes that make black holes likely are those that define how stars change and develop through their lifetimes.

We haven't been around long enough to watch a single star go through this process, which usually takes billions of years, but we've observed enough stars in different stages of their development to make it very likely that these theories for how black holes could form are correct.

Remarkably, the idea of special stars that don't let out light was dreamed up over two hundred years ago. John Michell, an astronomer and geologist from England born in 1724, was thinking one day about the concept of escape velocity—something that eventually would be a crucial factor for the space program. If you throw a ball in the air, it falls back down to Earth.

Thanks to Newton's work on the force of gravity, Michell knew that this is because the ball can't escape our planet's pull. Before it gets high enough, the downward acceleration from the Earth's gravity slows it to a stop and sends it falling back. If Superman had been dreamed up in Michell's day, however, he could have thrown a ball faster than 11.2 kilometers per second (twenty-five thousand miles per hour), which means it would have escaped before gravity dragged it back.

It might seem that this minimum speed limit would make it impossible to send a rocket into space. Anyone who has seen a launch from Cape Canaveral will know that space probes take off much slower than twenty-five thousand miles an hour—to begin with, they appear to crawl their way into the sky. But escaping from the Earth's pull is far easier than it sounds.

First, we can cheat a little by sending something off into space eastward near the equator, moving against the Earth's spin, which means we have to achieve only around 10.7 kilometers per second (twenty-four thousand miles an hour) because the rotation of the Earth gives the rocket a boost. But more important, the farther away from the Earth our rocket gets, the lower the escape velocity becomes. Because the rocket is constantly under power it can take off slowly, and as long as it keeps moving up, it will escape. As it rises, the escape velocity becomes lower and lower. If Superman throws a ball into space it has to have that escape velocity at the moment it leaves his hand, because nothing else can push the ball upward. The only force the ball experiences after being launched is the downward force of gravity, which is why it needs to start off with such a high velocity.

Michell imagined how escape velocity would vary if he were on a much bigger planet, or even a body as enormous as the Sun. Newton had shown that the force of gravity goes up with the mass of the planet or star—as the body you are standing on gets bigger, then the escape velocity increases too. What would happen, Michell wondered, if the mass was so great that the escape velocity was faster than the speed of light? Under those circumstances the light would never make it away from the star—no light would get out. It would appear to be a dark star even though it blazed furiously away on its surface. (Michell didn't call his concept a black hole— the name was dreamed up by American physicist John Wheeler as recently as 1969.)

No one took much notice of Michell's idea, published in the *Philosophical Transactions of the Royal Society* in 1783. It was a hypothetical concept, not much different in its philosophical abstraction from concerns about how many angels could dance on the head of a pin. (Apparently this was never a true medieval philosophical subject; it was dreamed up in modern times as an example of the kind of thing it was thought medieval scholars worried about.)

It wasn't until the early part of the twentieth century that anyone would come up with a way to envisage black holes with mathematical precision. Einstein's newly developed general theory of relativity predicted that light would be influenced by gravity. General relativity predicted that massive bodies, like stars, distorted the very fabric of space. As light traveled through the warped space, its path would be bent, sending the straight light beam around a corner.

This idea inspired German physicist Karl Schwarzschild to consider how larger and larger stars would warp the path of the light they emitted. It was 1916, and remarkably, Schwarzschild managed to come up with this concept while fighting in the First World War, using Einstein's equations to describe the action of a star on light using math. Of itself this was no surprise (except for Schwarzschild's amazing ability to undertake serious mathematical work on the battlefield)—but a strange possibility dropped out of the numbers. Just as Michell had found with his basic assumptions on escape velocity, Schwarzschild showed that a massive enough star would bend space so far that light would never get away. Instead it would turn in on itself and return into the star.

Schwarzschild thought this was nothing more than a mathematical nicety without a real application, because the ability to bend space was dependent on both the mass of the star and its size. It wasn't enough to have a supermassive star; it would also have to be much tinier than any star that had so far been observed. To get our Sun, which at 1.4 million kilometers (865,000 miles) in diameter is on the small side for a star (technically it's a yellow dwarf), into a state where its mass was concentrated enough to turn it into a black hole, you would have to compress it until it was just two miles across.

However, when Indian physicist Subrahmanyan Chandrasekhar and American Robert Oppenheimer, who would later head the Manhattan Project, looked into the practicalities of Schwarzschild's strange compact stars, they realized that there was a way for such compression to occur. Any star has a huge amount of mass—the Sun, for instance, is over 300,000 times as massive as the Earth. All that material in the star is pulling together with a vast gravitational attraction that is constantly trying to compress it like we would scrunch up a paper ball. While the star is very active, the outward pressure from the nuclear reactions that power it keeps the star "fluffed up," but as the nuclear fuel runs low, that pressure drops and the star begins to collapse.

Now another physical force comes into play—a quantum feature called the Pauli exclusion principle, which means that similar particles of matter that are close in distance must have different velocities. This requirement will counter the gravitational collapse as a star cools—unless the star is so massive that gravity overwhelms the Pauli effect. The mass required for this to happen is around one and a half times that of the Sun (so our star is not on its way to being a black hole).

Sometimes such a star explodes as a supernova, seeding the universe with heavy atoms. But if this fails to happen, the theory goes, the star should contract, getting smaller and smaller until the gravitational intensity is such that light never escapes—it has become a black hole. In theory, though, there is nothing to stop the contraction from continuing indefinitely until a singularity is formed, a point of infinite density, at the center of the black hole.

The outer surface of the black hole that we would see (or rather that we wouldn't see) isn't an actual material surface, like the outside of a normal star. The remnants of the star are much smaller than this boundary, known as the the event horizon. The surface of the black hole is just the radius at which the gravitational force becomes so strong that light cannot get out.

Cosmologists think that there are black holes out in space ranging in size from a couple of times the mass of the Sun to thousands of times that size for the supermassive black holes thought to be at the center of most galaxies. And in the normal way of forming a black hole, there is no way to make one from a star if it's smaller than that minimum radius of around one and a half times the Sun. But there is, in theory, another way that a black hole could be produced.

If you could cram together *any* amount of material with enough force—this would require much more energy than the gravitational pull experienced on a star—then you could, in theory, produce a black hole. Such a black hole could be formed not from a stellar mass, but from just a spoonful of matter—it could be as small as you liked. This is where the Large Hadron Collider at CERN comes in—and in some people's minds where it could put the world at risk. In theory, the LHC could come close to battering particles together with enough energy to make microscopic black holes pop into being.

To be precise, it's not so much the Large Hadron Collider that would be making the micro black holes, as the LHC with the addition of a little help from another universe. According to the best-supported current theory, even the vast energy levels the LHC can generate are not enough to bring tiny black holes into existence. But if, as some theories suggest, there are many parallel universes, and gravity can leak from one universe to another, it is thought that extra compressing force would be enough to enable the LHC to spawn micro black holes.

If they do appear, they will generate a lot of interest in the existence of other universes in near reach of our own. Apart from producing a fascinating subject of study in the black hole itself, the LHC would be providing fundamental data on the nature of the universe—this is part of the reason scientists get so excited about the collider, and believe that the immense amount of money spent on it is worthwhile.

So let's imagine the LHC really did make one of those microscopic black holes. We're talking about a body compressed to a size much smaller than an atom, not directly visible in any way. But in the Hollywood model of the black hole, this would be attracted toward the center of the Earth and would start to act something like the central character in the Pac-Man game, eating its way through the surrounding matter, absorbing it and growing ever bigger until it eventually swallowed the whole world. Here, surely, is good reason to worry about the Large Hadron Collider. If a micro black hole is formed, we'll see the world eaten away from under our feet.

Luckily, this picture misses one of the key points of the modern theory of black holes—something called Hawking radiation. Quantum theory tells us that in space, pairs of particles are constantly winking into existence. Usually these particles immediately annihilate each other and disappear again. It's a form of constant quantum interchange between energy and mass. But at the event horizon of a black hole, the tendency would be for one particle to be sucked into the black hole while the other shoots off into space, forming the Hawking radiation. It would give a black hole a faint glow around its perimeter.

About the most fundamental concept in all of physics is the conservation of energy. This now comes into play. The result of this interaction between the black hole and the particle pair is a net decrease in energy of the black hole. Every time it absorbs half a pair and gives off Hawking radiation, it loses energy. Effectively such a small black hole will fizzle out of existence long before it can eat anything up.

If micro black holes were formed (and remember they could come into being only if something like the gravitational field of a parallel universe gave a boost to the LHC's energy), they would disappear before they could be observed. All we would see is a little spatter of particles as the holes were transformed into Hawking radiation. Most physicists at CERN would love it if micro black holes were formed—but they would only ever be seen indirectly, lasting for a minuscule period of time. They would not provide any threat to our survival.

Though black holes are unlikely to do any damage, another worry of those who feel that the LHC could destroy us all is the formation of strangelets. We'll come back to what these are in moment. Although their existence involves the piling of hypothesis on hypothesis, if strangelets really were formed and behaved as some predict they would, the LHC could destroy the entire universe. Strangelets are hypothetical tiny chunks of strange matter, a form of matter different from the stuff we normally experience.

To understand what the threat from strangelets would be, we need to take a quick diversion into the nature of matter itself.

Most of us will remember from high school science that all matter is made up of three types of particle: protons, neutrons, and electrons. The protons and neutrons sit in a vibrating bundle in the tiny nucleus of the atom, while the electrons are distributed in a fuzzy cloud around the outside. These three types of particles used to be considered fundamental particles—ones that have no subcomponents—and electrons still are seen that way, but we now believe that neutrons and protons are made up of smaller particles called quarks.

Quarks were named by American physicist Murray Gell-Mann. He intended the word to be pronounced to rhyme with "dork," hearing the sound in his head before he worked out how to spell it. Soon after dreaming up the name, he came across the line "three quarks for Muster Mark!" in James Joyce's novel *Ulysses*. This sounded apt, as quarks often come in groups of three, but Gell-Mann wanted to keep his original pronunciation. Mostly he is ignored, and the word is pronounced to rhyme with "bark."

Quarks (however you pronounce them) come in different types, whimsically called "flavors," such as "up," "down," "charm," and "strange." These names are just labels—there is nothing inherently strange about a strange quark or charming about a charm quark (nor is anything pointing up or down in the flavors with those names). A proton is made up of two up quarks and one down, while the neutron is two downs and one up. But what if different quarks were used to build matter—what would be the result?

We have already met one alternative kind of matter based on a different assembly of quarks: antimatter. Every quark has an equivalent antiquark—and it's equally possible to make a kind of proton from two antiup quarks and an antidown quark. The result is an antiproton—just like a proton but with a negative charge instead of positive. As we have seen, it is also possible to have antiatoms, made up of antiprotons and antielectrons (also called positrons).

A more speculative construct is strange matter. Instead of being built from the familiar building-brick particles like protons and electrons, this would consist of a collection of quarks that aren't bound up in triplets, but would form a stable material that has roughly equal numbers of up, down, and strange quarks in it.

In principle, strange matter should be more stable than a normal atomic nucleus, and atomic nuclei should decay irreversibly to form strangelets—strange-matter particles—but the probabilities involved for any particular nucleus are so small that the chances are that none has yet had time to form in the lifetime of the universe.

If a strangelet did form and collided with a normal atomic nucleus, it could act as a catalyst, converting that nucleus to strange matter too. This process would give off energy, which could produce another strangelet, and so forth. The result would be a chain reaction, like the one powering a nuclear power station, but happening with ordinary matter at room temperature, resulting in an increasingly rapid breakdown of matter into strangelets.

An observer would see matter beginning to fragment into invisible particles, this decay spreading faster and faster until, in principle, the whole Earth had disappeared. The final result of this uncontrolled chain reaction could be that every speck of matter in the universe is converted into strangelets, almost as fundamental a transformation of existence as the big bang.

It sounds like scary stuff, but the CERN physicists point out that collisions of similar energy to those produced by the Large Hadron Collider are happening all the time in nature—they just don't take place under our control and observation, as they will be at CERN—yet we don't see matter breaking down into strangelets out in space. We also don't even know that strange matter *can* be formed—it has never been detected. It is also believed that strangelets would be stable only at very low temperatures, and that the probability of them forming drops as the energy of the collisions involved increases, so it's a strange worry to have for the LHC, when they haven't formed in lower-energy colliders. In the end, an awful lot of ifs have to be assembled to make the strange-matter scenario even vaguely possible.

Opponents of the Large Hadron Collider, led by German chemist Otto Rössler, asked the European Court of Human Rights to block the use of the LHC, but the court turned down the request. The court rejected their application for an emergency injunction but allowed the case to continue being argued, implicitly dismissing the opponents' argument, as the case would not have been heard until after the LHC was started up. As it happened, the LHC was crippled for over a year by a technical fault in September 2008 before it was fully switched on. It wasn't until October 2009 that particles began to travel around the accelerator again, and it was late November 2009 before the first collisions occurred. This delay made it possible for the lawsuit to continue, but the European legal attempts to block the LHC were not revived.

An equivalent action in the U.S. courts, brought by retired nuclear safety officer Walter Wagner and science writer Luis Sancho, was dismissed by District Judge Helen Gillmor in Hawaii in

September 2008. The judge ruled that the LHC did not fall under the court's jurisdiction, as the United States did not contribute enough to the project to bring it under domestic environmental regulations.

These concerned individuals are not mere attention seekers or cranks. They are people with a science background who are genuinely worried about what might happen when the Large Hadron Collider is fully operational. But the vast majority of scientists would argue that the worries expressed by the protesters are so hypothetical that they don't form a basis for stopping the research.

We can dream up hypothetical world-shattering consequences for almost anything we do. The LHC may be an extreme example, but even here, the chances are so small that arguably we should concentrate our efforts to avoid Armageddon elsewhere. There are more immediate and realistic threats of mass destruction.

Few indeed would argue about the reality of one such threat to the survival of the human race, arising from an activity that we have been engaged in since the 1940s. That's the use, and the terrifying misuse, of atomic power.

CHAPTER THREE
ATOMIC DEVASTATION

||||||||||||||||||||||||||||||||||||||||||||||||||||||||||||

> *If atomic bombs are to be added as new weapons to the arsenals*
> *of a warring world, or to the arsenals of nations preparing for war,*
> *then the time will come when mankind will curse the names of*
> *Los Alamos and of Hiroshima.*
> —Julius Robert Oppenheimer (1904–67), quoted in
> *Robert Oppenheimer: Letters and Recollections,*
> ed. Alice Kimball Smith and Charles Weiner (1980)

The Second World War, of all wars, was the one that was won by science. There was the cracking of the German and Japanese codes and the breaking of the Enigma cipher machines; the development of radar, enabling a whole new ability to detect attacking aircraft and ships before they were in visible range; and the introduction of a new science, operations research, which applied the techniques of math and physics to the battlefield—for example, using statistics to deduce the most deadly spread of depth charges. But most dramatically of all, science brought nuclear energy to the battlefield.

When I was a teenager in the 1970s, there was still a real feeling of living under threat, knowing that a nuclear attack could be on the horizon. While at university I gave serious consideration to moving to the remote Scottish islands once I had graduated, on the assumption that there would be a better chance of surviving a nuclear holocaust there. Atomic weapons threatened us with destruction on an unimaginable scale. True Armageddon. It was a feeling shared by many throughout the time of the cold war. Writer Sue Guiney describes her feelings as a child in Farmingdale, New York:

> [The cold war] was the backdrop of my childhood. All the fears, real and imagined, are still there inside me, almost on a cellular level. . . . As a seven-year-old, I remember having air raid drills where we were lined up in the school hallway and told to sit curled up facing the walls. I remember thinking that my small curved back would make a perfect target for a falling bomb, and I had nightmares about it for years.

The threat to the world that so terrified us then dated to a discovery made in the 1930s—the potential for destruction that lay in the nucleus of the atom. But we have to go further back, to 1909, to see how the existence of atomic nuclei was discovered, the first major step on the path to the atomic bomb.

Until then it had been thought that an atom was a ball of positive charge with negative charges distributed through it, like fruit in a pudding. But New Zealand–born physicist Ernest Rutherford and his team proved things were different. At the Cavendish Laboratory in Cambridge, England, Rutherford's assistants Hans Geiger and Ernest Marsden were using the decay of the natural radioactive element radium to produce alpha particles—heavy, positively charged particles—which were fired at a piece of gold foil to see how the atoms in the gold influenced the flight of the particles.

Unexpectedly, a few of the alpha particles bounced back. Rutherford commented that it was "as if you fired a 15-inch shell at a piece of tissue paper and it came back and hit you." The Cambridge team's discovery showed that the positive charge in an atom was concentrated in a small, dense nucleus. A very small nucleus indeed. If the atom were blown up to the size of a cathedral, the nucleus would be like a fly, buzzing around inside it.

Rutherford transformed our understanding of the atom's structure, but he was to make a remark about the practical significance of his discovery that showed he hadn't realized its potential. In 1933, in an interview with the *New York Herald Tribune,* he commented, "The energy produced by the breaking down of the atom is a very poor kind of thing. Anyone who expects a source of power from the transformation of these atoms is talking moonshine."

What Rutherford was referring to was the concept of nuclear fission. This was a process named after the biological process of fission, where a cell splits into two. Rutherford had first split the nucleus of an atom in 1917, again using alpha particles, but in this case using the gas nitrogen as the target. In 1932 his assistants John Cockroft and Ernest Walton had taken this a stage further, using artificially accelerated protons—the positively charged particles from an atomic nucleus—to smash into lithium, splitting it into two alpha particles. This was the "breaking down of the atom" Rutherford referred to. It took a lot of energy to smash the atom, so even though energy was released in the process, it didn't seem to have a lot of use.

What Rutherford had overlooked in his dismissal of nuclear power was that there was a way to make the process self-sustaining. The means to make his "moonshine" real was dreamed up by a Hungarian exile in London as he waited to cross the road. This was the young physicist Leo Szilard. He was staying at the Imperial Hotel in London's Russell Square, and had been held up by the traffic lights at the corner of the square, where it adjoins Southampton Row. Thinking of Rutherford's pronouncement as he waited for the light to change, Szilard was struck by a sudden idea that almost seemed prompted by the arrival of the green light.

"It suddenly occurred to me," he later said, "that if we could find an element which is split by neutrons, and which would emit two neutrons when it absorbed one neutron, such an element, if assembled in sufficiently large mass, could sustain a nuclear chain reaction." It was like getting interest on money in a bank account. You invested one neutron, and you got two neutrons back. Invest both those and you would get four—and so it went on. The process could be self-sustaining if kept in check, or could run away, doubling in its rate with every reaction.

The key concept at the heart of both the nuclear reactor and the atomic bomb had come to Leo Szilard in those few moments it took him to cross the road. (At least, this is the version he told years after the event. It wouldn't have been the first time a scientist had exaggerated the significance of a eureka moment.)

Unusually for a theoretical physicist of the era, Szilard took out a patent on his idea, which he sent into the patent office on March 12, 1934. At the time, he thought it most likely that the substance used to produce a chain reaction would be the element beryllium. This lightweight gray metal, element 4 in the periodic table, was already associated with the production of neutrons, as it was used in the discovery of those neutrally charged particles from the atomic nucleus.

With his friend and fellow Hungarian physicist Eugene Wigner, who was visiting London from his post at Princeton, Szilard decided to take a step into the lion's den and challenge Rutherford on the validity of his moonshine claim. After all, Rutherford had a laboratory capable of testing Szilard's ideas, something the theoretician had no way (or inclination) to do himself. Szilard intended to win Rutherford over with his brilliance, persuading the world-famous scientist that a young Hungarian upstart had got the better of him.

The result was a failure, not helped by a clash of personalities. The older, bluff Rutherford was not impressed by the way Szilard had taken out a patent on his idea. As far as the New Zealander was concerned, this wasn't what real scientists did—they shared information freely. Furthermore, Szilard messed up his chances in a mistaken attempt to play to Rutherford's enthusiasms. He knew that the older man had discovered the alpha particle, so rather than describe a chain reaction as Szilard had first envisaged it, using neutrons, he made up a similar process for Rutherford's benefit using the heavier alpha particles. Rutherford knew this wouldn't work, and threw out the idea without having a chance to think about a reaction happening with neutrons.

Szilard had conjured up the basic premise of the atomic chain reaction. Always the purest of theoreticians, he had little interest in the nuts and bolts of making such a reaction happen. That was why he had approached Rutherford. He had put the bottle on display, but he was unable to let the genie out. That was the responsibility of the team of the Austrian Lise Meitner and the German Otto Hahn.

In 1938, Meitner and Hahn picked up on work that had been done by the Italian-born Enrico Fermi on the splitting of uranium nuclei. Uranium is a naturally occurring radioactive substance that comes in two forms. By far the more common is uranium 238, while uranium 235, with three fewer neutrons in the nucleus, is the rarer and more unstable of the two.

Meitner and Hahn pointed out that when a uranium nucleus splits, it gives off between two and four neutrons, which go flying off with high energy and could collide with another uranium nucleus. The collisions should cause *those* nuclei to split as well, giving off more neutrons. The result would be a chain reaction, just as Szilard had postulated. If handled in a controlled way, this would enable useful production of power from the fission. When the nucleus splits, a small amount of the mass in the nucleus is converted to energy according to Einstein's iconic equation $E = mc^2$, where E is the energy, m the mass that is lost, and c the speed of light. This energy, added up over millions of splitting nuclei, would make for very usable power.

But there was another possibility that also occurred to Meitner and Hahn: if the reaction was allowed to run away without control. In principle that first split could generate at least two neutrons, each of which could split another nucleus, producing at least four neutrons, then going on to release eight neutrons in the next generation and so on. In a tiny amount of time, with this doubling, a vast number of atomic nuclei would be giving off energy—in this mode there would not be a neat source of steady energy for a power station but the sudden explosion of a phenomenal bomb.

Although science did not arrive at this concept until the 1930s, British fiction writer H. G. Wells had already speculated about the use of the power locked up within the atom as early as 1913. In his book *The World Set Free,* Wells describes the destruction caused by an imaginary weapon. He called it an "atomic bomb"—the first time those words had been seen in that combination. At the time, Wells's speculation was swept aside by the horrors of the First World War, which soon followed publication of his book. This "war to end all wars" was so horrendous even without such a weapon that Wells's ideas seemed unnecessarily far-fetched. But years later, the term that Wells had dreamed up would return as a very real threat.

Interestingly, in *The World Set Free,* we also see the first suggestions of "mutually assured destruction"—the idea that it is rational behavior for nations to refrain from making war if a country has the means to totally destroy its enemy, and the enemy has the same capability in return. The sad reality Wells depicts is one where that assurance isn't enough, though. It is only after a massive atomic war that the remnants of humanity can build a new and utopian society without the threat of war. Because they have experienced the true horror of nuclear warfare, they want it never to happen again. They know that the next time it would probably wipe out the human race entirely, and so they dismiss the use of war to settle any future political differences.

Producing an atomic bomb to rival Wells's fantasy would not prove easy (probably just as well for the continued security of the world). Meitner and Hahn's discovery pointed the way, though, and soon after, in 1939, a team at Princeton discovered a significant difference between the two different "flavors" of uranium.

Uranium 238—the more stable version with three extra neutrons in the nucleus—proved to be much worse at absorbing neutrons and subsequently splitting than was uranium 235. If you wanted uranium 238 to undergo fission, you needed slow-moving neutrons, giving the nuclei a chance to absorb them, whereas uranium 235 was capable of latching onto high-speed neutrons.

The vast majority of uranium came in the 238 variety—99.3 percent of a typical chunk of uranium when dug out of the ground would be of this type. This was fine for generating nuclear power, because you could slow down the neutrons using special materials; but the slow neutrons were no good for the "all-at-once" fission required for a bomb. A bomb making use of slow neutrons would fizzle along rather than explode. So if a bomb were to be made from uranium, it would need to be mostly uranium 235, enabling the formation of a chain reaction with fast neutrons.

This proved to be a nightmare problem for anyone attempting to make such a bomb. It isn't easy to distinguish between uranium 235 and uranium 238. There's no point trying to chemically refine the uranium to separate the two different varieties (known as isotopes). The chemical properties of an element are determined by the electrons in the atom, and both isotopes have the same number of electrons. It's only the number of neutrons in the nucleus, and hence the atomic weight, that differs. So to separate uranium 235, it was necessary to find a way to split off tiny amounts of a chemical with a very slightly different weight. It would take several years to discover a way to do this, proving one of the biggest difficulties faced by the atomic bomb project.

The first country to take the idea of a nuclear weapon seriously was Germany. In April 1939, the chemist Paul Harteck wrote to the German war office that nuclear fission would "probably make it possible to produce an explosive many orders of magnitude more powerful than the conventional ones. . . . That country which first makes use of it has an unsurpassable advantage over the others."

The possibility of making such weapons was next picked up by Winston Churchill in the United Kingdom, while in the United States in August 1939, a letter from Albert Einstein warning of the dangers nuclear fission posed, written at the encouragement of the father of the fission reaction, Leo Szilard, was sent to the authorities, but it seems not to have raised much interest.

Initially, the difficulties of separating enough uranium 235 to make a bomb seemed insuperable. But in June 1940, American physicists Edwin McMillan and Philip Abelson, working at the Berkeley Radiation Laboratory, wrote a paper that suggested an alternative approach that would avoid the need for separating the uranium isotopes. If uranium 238 can be encouraged to absorb a slow neutron in a reactor, it becomes the unstable isotope uranium 239. This undergoes a nuclear reaction called beta decay, where a neutron turns into a proton, giving off an electron in the process (for historical reasons, the electron is called a beta particle in such circumstances).

The result of this reaction is the production of a new element, one that doesn't exist in nature. This element was later called neptunium. But neptunium is also unstable and soon generates another electron, adding a second proton to the nucleus to become the element that would be named plutonium. This is a material that is as suitable for making a bomb as uranium 235. And because plutonium is chemically different from uranium, it is relatively easy to separate. Remarkably, the openly published Berkeley paper had shown the first step of how to use a nuclear reactor to make the principle ingredient of an atomic bomb.

Initially work on nuclear power outside of Germany was slow, but the science community knew that one of the greatest physicists of the 1930s, Werner Heisenberg, was working on the German nuclear project. Under pressure from the United Kingdom, where the threat of invasion was becoming more and more tangible as other European countries fell to the German army, the United States began a colossal program to beat the Germans in the race to produce nuclear weapons.

In the summer of 1941, an unparalleled, industrial-scale scientific venture began. Called the Manhattan Project after the location on Broadway of the Army Corps of Engineers' headquarters, it was an uneasy collaboration between the military and university scientists. Different subprojects included building the world's first nuclear reactor in order to produce plutonium; attempting to separate uranium 235 using different techniques like gaseous diffusion, electromagnetism, and high-speed centrifuges (picking out the very slight difference in mass); and working out the practicalities of assembling an atomic bomb—not a trivial task even if you have the materials.

The Manhattan Project's reactor was not the original birthplace of plutonium. The initial discovery of the element was the work of U.S. scientist Glenn Seaborg, a tireless producer of new elements, who made the first tiny sample of the material in February 1941 at the University of California, Berkeley. (Seaborg's and Berkeley's roles in adding to the list of elements would later be commemorated with the elements americium, californium, berkelium, and even seaborgium.)

It wasn't until August of the next year that Seaborg, by now transferred to Chicago for the Manhattan Project, would produce enough plutonium for anyone to be able to see it. According to Seaborg, such was the demand to take a look at this wonder element that, bearing in mind its rarity and the danger it posed, he instead showed people a test tube with diluted green ink in it. There is an echo here of the miracle of the liquefaction of the blood of Saint Januarius which is said to take place in Naples, Italy, every year. Many suspect that the glass vial supposedly containing the saint's blood in fact holds a suspension of an iron salt that gives the right effect. But to the believers it's the real thing, just as those who saw Seaborg's tube of ink believed they had been in the presence of a new, world-changing element.

On December 2, 1942, the first great breakthrough of the Manhattan Project occurred—at the drab location of a converted squash court in Chicago. The most modern piece of technology in the world was tucked away under the bleachers of a dusty, disused football stadium. Three years previously, the University of Chicago football team had been closed down by a college president who believed that the game was a distraction for those who should be concentrating on academic matters. Now the nineteenth-century structure of the stadium, all too like a mad scientist's retreat with its Gothic arches and grotesque statues, would house the ultimate mad scientist's dream.

Here, in a claustrophobic space under the stand, Fermi's team had assembled the world's first nuclear reactor—not built with any vision of producing cheap electricity, or for pure research. This was a structure intended only to produce plutonium for a bomb. It was referred to, almost sarcastically, as an "atomic pile" as it was constructed from a twenty-foot-high pile of carbon bricks, used as moderators to slow down the neutrons so that the six tons of uranium 238 scattered through those graphite blocks could interact with the particles. These slugs of uranium oxide were interlaced with so-called control rods made of cadmium, a material that absorbed neutrons much faster than uranium, so when the rods were in place, the reactor was damped down, preventing a runaway reaction from occurring.

Perched on a platform above this makeshift construction, three men stood ready, prepared to sacrifice their lives. If the reaction went out of control, their job was to douse the carbon bricks in neutron-absorbing chemicals. But by the time they had done this, they would already have received fatal doses of radiation. It's hardly surprising that their hopes for a safe test of the reactor were greater than those of any of the others present.

According to project mythology, Fermi insisted that the scientists and invited guests have a lunch break before proceeding with the first live test of the reactor. All but one of the control rods were withdrawn already. The final rod was pulled back and the Geiger counters monitoring radiation from the pile began to roar. The neutron count had shot up. The pile stayed critical—generating enough neutrons to make the reaction self-sustaining—for a little over four minutes before Fermi shut it down. The production of fuel that would eventually be at the heart of the first atomic bomb had been proved possible.

In March 1943, the most famous part of the complex Manhattan Project operation was brought online. Under newly appointed director Robert Oppenheimer, what had been a ranch school at Los Alamos in New Mexico was converted into a cross between a holiday camp, a university, and a factory, where arrays of mechanical calculators could churn through the complex arithmetic required to predict the behavior of a totally new kind of weapon, and where the raw materials being assembled elsewhere would be pulled together to build the first atomic bombs.

One of the difficulties that took up much of the time of the inhabitants of Los Alamos was working out just how to make a bomb based on nuclear fission explode, rather than generate energy in a slow and steady fashion. It is easier said than done to create a nuclear explosion. In a bomb, it's necessary to get the materials over critical mass, the amount required for the process to run away with itself, very quickly. The production of neutrons sparking the chain reaction has to peak suddenly. Otherwise, parts of the radioactive substance will go critical at different times and it will blow apart, yielding a tiny amount of its potential and leaving most of the radioactive matter intact.

Two different techniques would eventually be employed to achieve critical mass quickly enough. The bomb based on separated uranium 235 used the relatively simple gun method. Here a cylinder of uranium was shot into a hole in another piece of uranium at high speed, taking the combined mass suddenly above critical. The plutonium bomb used an alternative approach, where a hollow sphere of plutonium segments was forced in on itself by multiple simultaneous explosive blasts from different directions.

This second approach was much more complex, needing lengthy experimentation with the conventional explosive charges that would be used to drive the plutonium together. The different segments all had to arrive at the same time, requiring exquisitely precise coordination of a sphere had to be forced inward, using special shaped charges called explosive lenses that focused the blast of the conventional explosive in a particular direction. But because of differences between plutonium and uranium, the gun approach could not have achieved criticality quickly enough for plutonium, and the imploding sphere had to be attempted.

As it happened—though no one on the Allied side knew it—the race to complete an atomic weapon was one-sided. In part thanks to a successful raid on a Norwegian plant that produced a substance called heavy water, used in the German experimental reactors to slow down neutrons to enable them to work with uranium 238, the Germans didn't even succeed in getting a reactor working properly before the war came to an end. Some of the German scientists involved claimed after the war that they had intentionally worked as slowly as they could to prevent the Nazis from getting such a devastating weapon. This has been seen by some as an after-the-event defense, by others as true—we will never be sure.

Long before the bombs were ready, the great Danish physicist Niels Bohr argued for a form of deterrence. He believed it wasn't necessary to actually deploy the bombs, but merely to have the capability to use them. The devastating power of atomic weapons was such, he believed, that the sheer deterrent effect of what *might* be done with them should be enough to stop or prevent wars. To make the deterrent effective, it wasn't sensible for one country alone to push ahead, as this would inevitably result in an arms race—perhaps with the Soviet Union once the war was over. Instead, he believed, it was essential to share the technology widely, building up an international state of mutual trust. Few agreed with him.

By this time the first bombs were almost complete. The alternative techniques for achieving critical mass made them very different in appearance. It was originally thought that the uranium bomb, nicknamed "Thin Man" after the hugely popular Dashiell Hammett detective novel and movie, would be extremely long and thin—perhaps 0.6 meters (two feet) wide by 5.5 meters (seventeen feet) long. Most of this length was for the gun to shoot in the uranium slug. But the gun could be made shorter if the slug didn't need to move as fast—and it was discovered that by shielding the bomb from natural stray neutrons, the uranium would have less tendency to explode prematurely. This meant the bomb makers could take the length down to around two meters (six feet), and the bomb was rechristened "Little Boy."

No such slimming was possible for the plutonium bomb. The need for an inner spherical casing containing the charges that would implode the segments of plutonium meant that "Fat Man," as it was called, was never going to get much slimmer than the initial estimates of three meters by 1.5 (nine feet by five). Because of its complex mechanism, the plutonium bomb was less likely to work, but it still had to be attempted. With larger-scale reactors up and running, plutonium was relatively easy to make. The plants working on separating uranium 235 from uranium 238 were proving slow, and it was feared that only one uranium bomb could be made by early 1945, when it was thought it might be necessary to use nuclear weapons. The plutonium bomb had to be brought to completion as well—and that meant a test of that tricky implosion technique.

A site for the test had been arranged at the Air Force bombing range at Alamogordo, New Mexico. At least, that's how the location is normally described. But the White Sands range is enormous, the size of a county. The site was a good sixty miles from Alamogordo, tucked away behind Oscura Peak.

The test was given the evocative code name "Trinity" by Oppenheimer, whose poetic leanings included a knowledge of John Donne's *Holy Sonnet* 14, which begins:

> *Batter my heart, three-person'd God; for You*
> *As yet but knock, breathe, shine, and seek to mend;*
> *That I may rise, and stand, o'erthrow me, and bend*
> *Your force to break, blow, burn, and make me new.*

Thus, Oppenheimer associated the Holy Trinity with the kind of impact it was imagined the bomb would have. That's the usual story for the origins of the name, though Oppenheimer was less certain himself that this is where the label came from. He said in a letter that he wasn't sure why he had used "Trinity." He *did* have a Donne poem in his mind at the time—but it was another one that makes no reference to the Trinity. He just guessed that the one Donne poem triggered an association with another, but it's a pretty slim connection.

The test bomb was perched on top of a 33 meters (110-foot)-high metal tower to simulate an explosion in the air as it dropped from a plane. There it sat for over thirteen hours through a thunderstorm, as the nervous team, ensconced in a concrete command center some five and a half miles away, worried about lightning strikes and the potential for the unstable weather system to carry radioactive material from the explosion into inhabited areas.

The guests and observers not directly involved in controlling the test were sited on a hill twenty miles away. They had been warned of the delay and so did not arrive until 2 a.m., ten hours after the explosion was originally due to take place. As it was, it was just before 5:30 a.m. on July 16, 1945—after an uncomfortable wait in spartan conditions—that those observers witnessed the event that was to bring Armageddon closer to a man-made possibility. The young scientist Joe McKibben switched the bomb to automatic timer forty-five seconds before detonation. Warning rockets flashed from the camp to alert those around.

In a tiny fraction of a second after the timer kicked in, the explosive lenses around the plutonium sphere sent a spherical shock wave pushing the heavy metal inward, clamping the radioactive source together into a supercritical mass. The Fat Man plutonium bomb at the Trinity site exploded, with a visual impact that would become the template for movie-effects designers for decades to come. Physicist Otto Frisch, nephew of Lise Meitner, the codiscoverer of nuclear fission, described what he experienced:

> And then without a sound, the sun was shining; or so it looked. . . . I turned round [to avoid being blinded, observers had their backs to the explosion initially], but that object on the horizon which looked like a small sun was still too bright to look at. . . . It was an awesome spectacle; anybody who has ever seen an atomic explosion will never forget it. And all in complete silence; the bang came minutes later, quite loud though I had plugged my ears, and followed by a long rumble like heavy traffic very far away. I can still hear it.

Another physicist who witnessed the test, Isidor Rabi, spoke of the aftermath:

> Finally it was over, diminishing, and we looked toward the place where the bomb had been; there was an enormous ball of fire which grew and grew and it rolled as it grew; it went up into the air, in yellow flashes, and into scarlet and green. It looked menacing.

Within days, materials for the Little Boy uranium bomb, and the physicists to assemble it, had arrived on Tinian Island in the Pacific, from where bombers based at the North Field base were already pounding Japanese cities with conventional weapons on a daily basis. The intention had been to drop the Little Boy bomb on August 1, but the weather prevented it. Instead, the world's deployed nuclear arsenal was doubled in size the next day, when a Fat Man plutonium bomb arrived, confidently assembled after the Trinity test.

Days ticked by. Paralleling the early hours of the birth of the first atomic explosion, it was at 2:45 a.m. on August 6, 1945, that the B-29 bomber Victor 82 took off. It is better remembered now as the *Enola Gay*, named after the mother of pilot Colonel Paul Tibbets, onetime personal pilot of General Dwight D. Eisenhower. Tibbets headed up the 509th Composite Group, the air combat group assembled with the sole task of dropping the atomic bomb.

The Little Boy uranium bomb was dropped from the *Enola Gay* over Hiroshima six and a half hours later, dispatched from a height of thirty-one thousand feet. The people of the city were expecting an attack. After the recent terrible firebombing of Tokyo, they had been preparing. On that August day, crowds of schoolchildren were working hard making firebreaks to reduce the spread of fire in case incendiaries were dropped as they had been on Tokyo. It was hot and humid, but the work was necessary.

There were, no doubt, some wary looks when the *Enola Gay* accompanied by its two observer aircraft flew overhead—but the three planes would not have been considered much of a threat. They were clearly not part of a major bombing raid. And then that single payload fell.

From the flight deck of the *Enola Gay* it seemed an unreal experience. Pilot Tibbets later commented:

> When I level out [after dropping the bomb] the nose is a little bit high and as I look up there the whole sky is lit up in the prettiest blues and pinks I've ever seen in my life. It was just great.
>
> I tell people I tasted it. "Well," they say, "what do you mean?" When I was a child, if you had a cavity in your tooth the dentist put some mixture of some cotton or whatever it was and lead into your teeth and pounded them in with a hammer. I learned that if I had a spoon of ice cream and touched one of those teeth I got this electrolysis and I got the taste of lead out of it. And I knew right away what it was.

The bomb exploded. With a temperature at its core of around 60 million degrees Celsius, far hotter than the surface of the Sun, the initial flash vaporized some individuals and turned everyone openly exposed to it for around half a mile around into a fused carbon relic. Immediately after came the shock wave, which combined with that initial flash killed around seventy thousand people—flattening nearly as many buildings in the process.

Above ground zero, a vast mushroom-shaped cloud—soon to be the definitive marker of nuclear explosions—rose into the stratosphere. In reality, this shape of cloud had nothing to do with the Hiroshima blast being a nuclear explosion—a large conventional explosion will also generate a mushroom cloud—but the association of photographs from Japan and from later tests with atomic weapons is fixed forever in the human psyche.

The White House rushed out a press release containing a statement by the president.

> Sixteen hours ago an American airplane dropped one bomb on Hiroshima, an important Japanese Army base. That bomb had more power than 20,000 tons of T.N.T. It had more than two thousand times the blast power of the British "Grand Slam" which is the largest bomb ever yet used in the history of warfare. . . .
>
> It is an atomic bomb. It is a harnessing of the basic power of the universe. The force from which the sun draws its power has been loosed against those who brought war to the Far East. . . .
>
> The battle of the laboratories held fateful risks for us as well as the battles of the air, land and sea, and we have now won the battle of the laboratories as we have won the other battles. . . .
>
> We have spent two billion dollars on the greatest scientific gamble in history—we won.

The press release went on to make it clear that Hiroshima might be only the beginning. Japan should be ready for a true holocaust.

> We are now prepared to obliterate more rapidly and completely every productive enterprise the Japanese have above ground in any city. We shall destroy their docks, their factories, and their communications. Let there be no mistake; we shall completely destroy Japan's power to make war.

The new bomb was, the president was making clear, much more than just another weapon.

The immediate impact of the Hiroshima bomb was no greater than the effects of the vast firebombing raids that had already been undertaken by the Allies. These raids were themselves vehicles of mass destruction, first deployed on the night of July 27, 1943, over Hamburg in Germany. Such was the massive scale of the firebomb attack that the heat blasted air up into the atmosphere, causing devastating winds that snapped trees like twigs. Road surfaces and the glass in windows melted in the intense heat. It was, the survivors said, a firestorm (*Feuersturm* in German). Around 50,000 people died that night.

Horrific though the Hamburg raid was, it was to be repeated elsewhere, notably Dresden and Toyko, with even more casualties after British and American scientists carefully studied just how to deploy their firebombs to ensure the most destructive effect. But Hiroshima was different. First, it was the result of a single weapon, rather than a mass raid by wave after wave of bombers. And more terrifying still, the effects of the attack on Hiroshima were not over once the dust had settled.

Death did not cease in the minutes following the initial explosion. Radiation sickness would nearly double the initial death toll to well over 100,000 people during the subsequent year. That was out of a population of 350,000. Some have put the eventual total at double this. Although the explosive power of nuclear bombs is terrible, this is the truly terrifying aspect of such weapons: the silent, invisible, deadly action of radioactivity. From the very first days of working with radioactive materials, it became clear that the remarkable power of these elements was a two-edged sword. One of the earliest radioactive elements discovered was radium—and this element would claim the life of its discoverer, the double Nobel Prize winner Marie Curie, born Maria Sklodowska.

Working with her husband, Pierre, Marie Curie was studying pitchblende, a mineral found in north Bohemia that contained uranium. Pitchblende was mined near what's now Jáchymov in the Czech Republic, and after uranium had been extracted from the ore to be used to color pottery glazes and tint photographs, the residual slag was dumped in a nearby forest. Without the uranium, the pitchblende proved still to be radioactive—in fact, whatever the other radioactive material in pitchblende was, it was much more radioactive than the uranium itself.

Marie Curie wrote to her sister Bronia that "the radiation that I couldn't explain comes from a new chemical element. The element is there and I've got to find it! We are sure!" After working through tons of the pitchblende slag, painstakingly processing the material by hand, the Curies identified two new elements in the remaining material—polonium and radium. Radium, named for the Latin for "ray," was finally isolated in 1902 in its pure metal form and proved to be the most radioactive natural substance ever discovered.

Although Marie Curie lived until 1934, her death at the age of sixty-seven from aplastic anemia is almost certainly due to her exposure to radioactive materials, particularly radium. To this day her notebooks and papers have to be kept in lead-lined boxes and handled with protective clothing, as they remain radioactive. Marie Curie was the first victim, but radium would go on to kill others. It was seen in the early days after its discovery as a source of energy and brightness, it was included in toothpastes and patent medicines—it was even rubbed into the scalp as a hair restorer.

The application of radium that would bring it notoriety and would emphasize the dangers of radioactivity was its use in glow-in-the-dark paint. Unlike modern luminous paint, which has to be activated by bright light, storing up energy to release later, radium glows constantly from its natural radioactive energy. It was frequently used to provide luminous readouts on clocks and watches, aircraft switches and instrument dials; its eerie blue glow was seen as a harmless, practical source of nighttime illumination. It was only when a number of the workers who painted the luminous dials began to suffer from sores, anemia, and cancers around the mouth that it was realized that something was horribly wrong. The women workers would regularly bring their paintbrushes to a point by licking them. This left enough radioactive residue in their mouths to cause cell damage. Eventually more than a hundred of the workers would die from the effects.

These weren't the first deaths from radiation. It has happened as long as human beings have been around, without realizing the cause. Natural radiation levels vary hugely from place to place.

Denver, for instance, has a significantly higher background radiation level than New York. Some rock types, notably granite, release significantly more radiation than others, particularly when naturally released radon gas builds up in houses. Throughout history, background radiation has triggered cancer in handfuls of individuals. But this is as nothing compared with the aftermath of the dropping of Little Boy.

Three days after Hiroshima, confusion raged in Japan. Some argued that, disastrous though the attack had been, it was a one-off that changed nothing. Against this backdrop, another B-29, named *Bock's Car,* with Major Charles Sweeney at the controls, took off from Tinian Island to drop the world's second plutonium bomb on the Japanese military arsenal at Kokura. Conditions were difficult. The plane came under antiaircraft fire, and visibility was poor because of palls of smoke from conventional bombing attacks. Major Sweeney diverted to the alternate target he had been given—the city of Nagasaki.

With a yield of nearly twice that of the Little Boy, the Fat Man bomb might have been even more devastating, but the terrain around Nagasaki contained the blast to some degree, reducing the impact on the outskirts of the city—even so, another seventy thousand or more died immediately as the bomb exploded. The Japanese realized that Hiroshima had been no one-off, last-ditch attempt by their enemy. The Japanese surrender followed soon after.

During the Second World War, the Soviet Union had begun a limited nuclear program, but the intense pressures the Soviets faced in their battles with Germany meant that limited resources could be deployed. Now, as the war came to an end, two factors contributed to the Soviet Union quickly playing catch-up. One was espionage. During the war, a number of scientists with Communist sympathies, who believed it was essential to have a nuclear balance, were passing as many details as they could of both reactors and bombs to the Soviets. The other factor was a rapid buildup of effort, making use of as many materials as could be retrieved from the German nuclear program in the eastern section of occupied Germany.

Stalin was determined that the Soviet Union would not be held for ransom by American power, a possibility that he believed had been demonstrated in Hiroshima and Nagasaki. The nuclear program was officially made a state priority just eleven days after the bombing of Nagasaki. Its importance was emphasized by Stalin's putting his feared lieutenant and secret police chief Lavrenty Beria in charge of the development.

Stalin did not believe that the United States would use atomic weapons against the Soviet Union for anything less than a response to a direct attack—but he still felt that the USSR's having its own nuclear weapons was essential to maintain a balance of power. By August 1949, newly built Soviet reactors had produced enough plutonium to test a bomb that had a more than accidental resemblance to the Fat Man design. At 7 a.m. on August 29, Beria and his team witnessed the first Soviet nuclear explosion in a test in the remote regions of Kazakhstan. Although the USSR would never catch up with the United States in numbers of warheads, it would not be long before both superpowers had nuclear arsenals capable of immense destruction.

Did the bomb have to be used in Japan? It's too late to say now, but a committee of the scientists working on nuclear weapons development, meeting in June 1945 before the Trinity test took place, believed it wasn't necessary to use the atomic bomb in anger to have a deterrent effect. In the Franck report, named after committee chairman James Franck, the group argued against nuclear proliferation. If anything, the report suggested, stockpiling more and more nuclear weapons might be the trigger for an attack rather than a means of defense. "Just because a potential enemy will be afraid of being outnumbered and outgunned," the report commented, "the temptation for him may be overwhelming to attempt a sudden unprovoked blow."

The Franck report argued that the best approach for world safety would be to demonstrate the new weapon in front of an assembled group of United Nations observers, and for the United States to express its magnanimity by saying, "Look what we could have done"—but instead of using the bomb, inviting the rest of the world to share in the knowledge of nuclear technology so that everyone could work together to prevent the proliferation and the use of nuclear weapons.

Presciently, the report went on, "If the United States would be the first to release this new means of indiscriminate destruction upon mankind, she would sacrifice public support throughout the world, precipitating the race of armaments, and prejudice the possibility of reaching international agreement on the future control of such weapons."

We can't tell how Japan would have reacted to such a show of strength without actually experiencing the impact of an atomic bomb on a Japanese city—but we do know that proliferation and difficulties of agreement over arms control did follow from the use of nuclear weapons. As soon as the bomb was dropped, the arms race was on, though for a brief moment toward the end of 1945, it had looked as if such action would prove unnecessary.

The three leaders of the Western powers involved in the development of the successful bomb—President Harry Truman for the United States, and Prime Ministers Clement Attlee and William King for the United Kingdom and Canada—met in Washington, D.C., on November 11, 1945. The three men came up with a proposal that seemed to carry forward the spirit of the Franck report. In it they said that the atomic bomb was a means of destruction previously unknown to mankind, against which there could be no adequate military defense, and which could not be a monopoly for any nation.

To ensure that this was the case, the triumvirate suggested setting up a United Nations commission to eliminate "the use of atomic energy for destructive purposes" and to promote its "widest use for industrial and humanitarian purposes." This was agreed to by the Soviet Union, and in January 1946 the UN Atomic Energy Commission was established with a purpose that included "the elimination from national armaments of atomic weapons and of all other major weapons adaptable for mass destruction."

President Truman had a U.S. committee established to lay the groundwork for taking the concept of worldwide nuclear control from idea to practical plan. This group, relying heavily on the technical advice of Robert Oppenheimer, decided that the UN organization's mandate did not go far enough. The resultant Acheson-Lilienthal report, named for Dean Acheson and David Lilienthal, the chairmen of the committee and its board of consultants, proposed nothing less than handing every bit of the atomic behemoth, from uranium mines to nuclear reactors, over to a single, peaceful world organization.

This was a brief moment when, despite the two bombs dropped over Japan, the world could have stepped back from nuclear brinkmanship. The Acheson-Lilienthal proposals would have turned the atomic arms race on its head. Instead of nations competing to have the greatest destructive force, this American proposal for world peace suggested that nuclear technology should be spread to every country, along with the scientific and industrial capability to handle it. This technology would be controlled by the UN-owned organization, not by any nation-state.

Any attempt to subvert that central control would be deterred by the awareness that the rest of the world would respond by switching to the production of atomic weapons. This would be mutually assured deterrence, but one where any offender was inherently outnumbered—and the deterrence was at arm's length, as the atomic weapons did not yet exist, only the capability to produce them.

This idyllic world peacemaking body was never to be made a reality. Diplomatic relations between the USSR and the West were worsening. In March 1946, Winston Churchill, the wartime British prime minister, who still held a lot of influence and was dogmatically opposed to any collaboration with the Soviet Union, gave a speech in Fulton, Missouri, where the words "iron curtain" were first used to describe the division between the Soviet sphere of influence and the West.

Churchill took the opportunity of this speech to rubbish the suggestions raised in the Acheson-Lilienthal report. It would be wrong and imprudent, he said, to give the knowledge of the atomic bomb to the UN, and "criminal madness to cast it adrift in this agitated and un-united world." Churchill, who had an unwavering enthusiasm for maintaining secrecy, believed wrongly that the secret of atomic power could be kept away from the USSR. The Acheson-Lilienthal report was fatally weakened before it had a chance to deliver any results.

The possibilities for world control of nuclear power took a second blow with the appointment of the seventy-five-year-old archconservative financier Bernard Baruch to lead the U.S. delegation to the UN Atomic Energy Commission. Baruch took an instant dislike to the Acheson-Lilienthal report and rapidly moved to replace it with a plan that was more combative. Yes, Baruch said, it was possible to dismantle atomic weapons—but only when an adequate system of world control was in place, including punishments for those who violated the controls, and only at the behest of the relevant national governments.

Further splintering of atomic cooperation came when, in August 1946, President Truman signed the Atomic Energy Act. This was a domestic U.S. bill that closed the door on cooperation with the United Kingdom and Canada. With controls falling apart and international cooperation failing, each of the countries with the technological capability to make nuclear weapons went its own way. There was not to be another opportunity to move the world toward safety, away from the construction of opposing nuclear arsenals.

Instead, the creation of the new atomic arsenal led to plans in the United States that were much more hawkish. The military advice was to prepare for a total war that would be on a scale that dwarfed even the two world wars. A plan was drawn up to be prepared for a preemptive strike on the Soviets, hitting sixty-six cities with a total of 466 atomic bombs—bombs that didn't yet exist, but

soon would. What was being contemplated was the total oblitera-
tion of the enemy before it had a chance to strike.

The death and destruction at Hiroshima and Nagasaki were
awful in the true meaning of the word. Yet to support the oblitera-
tion plans, within months the United States was working on an
alternative weapon that was even more fearsome than the atomic
bomb. When nuclear weapons were first proposed, there was a
serious concern that the explosion would be so hot that it would
set the atmosphere on fire, fusing the nitrogen molecules that
make up the majority of the atmosphere. Although this was proved
an impossible outcome long before the first test, there was another
possibility arising from that immense heat that could be put to
use, according to a theory developed by Hungarian-born American physicist Edward Teller.

The original idea had come from Enrico Fermi. He speculated
that the intense heat of the fission explosion could be enough to
cause small molecules of deuterium (an isotope of hydrogen with
an extra neutron in the atomic nucleus) to fuse together. In the
process, the molecules would give off energy. This thermonuclear
reaction, nuclear fusion, is the working mechanism of the Sun. If
the deuterium could be made hot enough, there would be no need
to worry about the fussy chain reactions that were required for a
conventional atomic weapon—the fusion would continue until it
had worked through the fuel. The explosion had the potential to be
vastly more powerful than that of a basic fission weapon.

To give an idea of the scale of the explosion that nuclear fusion
could provide, the atomic bombs dropped on Japan at the end of
the Second World War were the equivalent of around twenty

thousand tons of TNT. It would only take twelve kilograms (twenty-six pounds) of deuterium to be exploded in a thermonuclear device to produce the equivalent explosive power of 1 million tons of TNT.

When Teller first got involved with this idea of an ultrapowerful bomb, alongside American colleague Emil Konopinski, his aim was to show that it wasn't possible. He wanted to eliminate this wacky idea of Fermi's. This was in 1942, when any form of atomic bomb was still a distant hope. But the more Teller worked on the idea, the more realistic it seemed to him. Such a thermonuclear device—a superbomb, or just "Super" as it was referred to—should be practical once a fission device existed to trigger it.

Teller became obsessed with the idea, and began to press for work on the Super, but he was quietly sidelined. There were enough problems getting a basic nuclear weapon to operate without building a device where the atom bomb was nothing more than a trigger. But Teller did not let the matter drop, and after the end of the war he was able to press forward the concept that would become known as the H-bomb or hydrogen bomb, even though in practice it would not be fueled with regular hydrogen.

Opinions were divided in the United States. Some felt that it was necessary to go the next step, now that the USSR had nuclear weapons. Others argued that the technical difficulties in building a thermonuclear weapon would be extreme, or that building such a weapon raised moral hazards. In a report on the hydrogen bomb from the U.S. General Advisory Committee (the body that advised the government on nuclear matters) there was a majority annex that spelled out this moral argument in no uncertain terms:

> We base our recommendation on our belief that the extreme dangers to mankind inherent in the proposal wholly outweigh any military advantage that could come from this development. Let it be clearly realized that this is a super weapon; it is in a totally different category from an atomic bomb. The reason for developing such super bombs would be to have the capacity to devastate a vast area with a single bomb. Its use would involve a decision to slaughter a vast number of civilians. . . . If super bombs will work at all, there is no inherent limit in the destructive power that might be attained with them. Therefore, a super bomb might become a weapon of genocide.

In a minority annex, physicists Enrico Fermi and I. I. Rabi went even further:

> The fact that no limits exist to the destructiveness of this weapon makes its very existence and the knowledge of its construction a danger to humanity as a whole. It is necessarily an evil thing considered in any light.

The committee hoped that the fusion bomb would never be produced, setting a limit to the totality of war, while Fermi and Rabi suggested inviting the nations of the world to solemnly pledge not to proceed with the development or construction of such weapons.

The U.S. administration was not in the mood for listening to such a viewpoint. The prevailing feeling was that to set aside the possibility of thermonuclear weapons amounted to a voluntary weakening of the military capability of the United States. It was thought that enemy countries, particularly the USSR, would see this as weakness, and that it would prove a trigger for opposing forces to align against America. The war was still fresh and raw enough to make even as successful and powerful a country as the United States wary of what it might face.

The race to take nuclear weapons to a further level of destructiveness had begun. On January 31, 1950, President Truman made a broadcast to the nation, announcing that he had directed the Atomic Energy Commission to continue work on atomic weapons, "including the so-called hydrogen or super bomb." Not only was the H-bomb to be built; Truman made sure that the rest of the world was well aware of it, and of America's nuclear supremacy.

There was one small problem with Truman's announcement. Although the basic concept of the thermonuclear device was sound, no one really knew how to make one work. It would require an immense explosion, well beyond the capabilities of the atomic bombs at the time, to heat a material like hydrogen sufficiently to make it fuse. Even the temperatures in the heart of the Sun aren't enough for this—it also takes the pressure exerted by the Sun's huge mass, and a quantum mechanical mechanism called tunneling, for the Sun to be able to fuse hydrogen.

The breakthrough, which was made around the same time as President Truman's statement, came from two scientists, Stanislaw Ulam and the ubiquitous Edward Teller. Instead of simply heating the fusion material, the radiation from the triggering fission device could be used to compress the material, using emissions from the explosion to compact the molecules closer and closer together, making fusion easier to initiate.

After months of technical testing of components, the first thermonuclear bomb was ready to be tried out at a remote island location, Elugelab on the Eniwetok Atoll in the South Pacific. Like the innocently named Little Boy and Fat Man, the bomb had a nickname—"the Sausage"—because of its long cylindrical shape.

When the bomb exploded on November 1, 1952, it produced an explosion with a power of over ten megatons—nearly five hundred times the destructive power of the Nagasaki explosion, totally destroying the tiny island. This was very much a test device—weighing over eighty tons and requiring a structure around fifteen meters (fifty feet) high to support it, meaning that it could never have been deployed—but it proved, all too well, the reality of the thermonuclear weapon.

Such was the bad feeling by the time of the test between Edward Teller and those involved in constructing the bomb—Teller probably felt that he wasn't being given the central role that he deserved—that Teller was not present for the test. Instead, he waited in the Geology Department of the University of California, Berkeley, making use of one of their seismographs to observe the moment that his baby came to life. He sent a telegram to Los Alamos that seemed to reinforce the fatherlike relationship he felt to the bomb. "It's a boy," it read. But any sense of the H-bomb giving the United States an unbeatable lead in security was short-lived. The Soviet Union followed less than a year later with its own thermonuclear device.

By 1954, the United States was ready for another test of a hydrogen bomb, at a location that would be remembered by the general public as a result of the piece of clothing named after it, long after details of the test itself dropped out of the news. It took place on the Bikini Atoll. This was a much more practical-sized bomb, hence its nickname, "the Shrimp." Though not at this stage a true self-contained weapon—it was more like the messy patchwork weapon used in the Trinity test—it required only a final housing to be able to be dropped from an aircraft.

This test proved that despite the deployment of the new electronic computers that were being used to make the necessary calculations—a huge advance on the electromechanical calculators used in painstaking number crunching for the Manhattan Project—the numbers involved were fiendishly difficult to get right. The scientists were expecting the bomb to be the equivalent of five megatons of TNT—big enough at 250 times the Japanese bombs. But in practice, the explosion was more like fifteen megatons and threw out debris far beyond the expected zone.

It was soon discovered after the Bikini test that the hydrogen bomb had a second deadly output over and above the impact of the original fission bombs—fallout. A mixture of debris from the target area where the bomb was exploded and uranium and plutonium from the bomb itself were flung high into the sky and rained down as a shower of deadly radioactive material. This fallout can spread for many miles around the explosion site, particularly if carried by the wind, vastly expanding the danger of radiation sickness and death.

This was demonstrated all too well in the days following the Bikini test, where eighteen thousand square kilometers (seven thousand square miles) would eventually receive significant contamination. Ninety miles away from the epicenter, on the Pacific island of Rongelap, strange white ash began to fall from the sky around four hours after the explosion, which had sounded like thunder to the occupants. They had not been warned of the test, and it took two further days before an evacuation of the island began.

In that time, the islanders began to develop symptoms of radiation poisoning from that seemingly harmless snowlike deposit that had formed on the island. In a typical response to radiation, ugly burns appeared on their skin. The victims began to vomit and suffer from diarrhea. Hair came out in chunks, and some islanders were coughing up blood.

For many, the hydrogen bomb was the ultimate example of humanity's destructiveness—a horrible and frightening deliverer of overkill. Teller, however, remained in love with the concept that had won him over in 1942, and for the rest of his life pressed for applications of thermonuclear devices for everything from mining to diverting asteroids that were on a collision course with Earth. He infamously dismissed concerns about fallout from the tests of hydrogen bombs, commenting that the risk was no worse than being "an ounce overweight." Charitably, we can assume he didn't know about Rongelap.

In the arms race that followed the first successful thermonuclear explosion, bigger and bigger bombs were developed. This demonstrated the truly terrifying nature of fusion devices—there is no theoretical limit to the size of explosion they can deliver. The largest ever exploded was a Soviet bomb tested on October 30, 1961, which was the equivalent of 58 million tons of TNT. Vast stocks of nuclear weapons were built up. The U.S. military had not forgotten Pearl Harbor, and assumed that an unprovoked attack from the Soviet Union could result in the majority of its weapons being destroyed. For this reason it stockpiled one hundred times as many weapons as were considered necessary to wipe out America's enemies. It was hoped that if the Soviets knew that there would always be enough weapons to destroy them, they would never make that first strike.

By the late 1950s Britain had also successfully tested a thermonuclear weapon, and China and France were to follow. Other nations such as India, Pakistan, and Israel have followed in the nuclear route, though Israel's stock is limited to fission weapons, and it is not clear exactly what capability India and Pakistan have available. Thousands of nuclear weapons have been accumulated.

Even today, fission bombs have not gone out of fashion, as thermonuclear devices will always be relatively unwieldy, but the more recent fission weapons are something of a hybrid. Conventional fission weapons of the kind used at Hiroshima and Nagasaki have a practical limit of around fifty kilotons—two to three times the power of the Japanese bombs—but this can be enhanced by a factor of ten using a technique known as boosting.

The fission chain reaction is limited by the number of neutrons of the right speed that can be pumped into the fissile material. In a boosted weapon, the standard neutrons from the fission process are added to by neutrons being generated in a small fusion reaction. A small amount of fusible material like deuterium is injected into the heart of the bomb. This undergoes nuclear fusion in the explosion, generating extra neutrons to boost the fission reaction. Because it's only a small amount; there is no need for all the sophisticated mechanism necessary to get a true fusion bomb to work—this is just a turbocharger for the conventional fission reaction. Where a modern atomic stockpile contains fission weapons, they are like to be such boosted devices.

There are inevitably a range of reasons why states have entered the nuclear club. For some, like Israel, it seems to have been as a result of perceived threat. As a small country, surrounded by enemies, it seemed to the Israeli state that the deterrent power of nuclear weapons was necessary—although the continued terrorist attacks since Israel acquired its nuclear capability seem to emphasize that nuclear weapons deter only states themselves capable of making nuclear attacks, as those attacking Israel have assumed that a conventional attack will not produce nuclear retaliation.

Others, like North Korea, seem to be aiming more for status than to deal with any perceived threat. The term "nuclear club" has a suggestion of us and them, the elite and the hoi polloi. It's easy from within a country that has a nuclear capability to dismiss this need for status as insignificant, but there is a real sense of "Why should they have it when we don't?" It's no coincidence that the permanent members of the UN Security Council all have nuclear weapons. Membership in the club is a badge of power.

At the time of writing, as the aftermath of the 2008 recession continues to have an effect, many in countries like Britain and France are questioning whether it is financially acceptable to maintain a nuclear deterrent, particularly where conventional military forces have been stretched by actions in Iraq and Afghanistan. Britain, particularly, has had huge debates over whether it should renew the Trident nuclear submarine program. Yet financial arguments about military benefits miss the point. Maintaining a nuclear capability is not primarily a military action, but a political one. Having nuclear weapons is about international prestige, about being *someone* in the family of nations.

Arguably this is mad politics rather than mad science. Once

the science was established, there was a clear opportunity for the politicians to control atomic power for worldwide good—but political fear or political necessity (depending on your viewpoint) made this dream impossible to realize. Despite the reduction in tension since the fall of the Soviet Union, and the significant reductions in numbers of weapons under the Strategic Offensive Reductions Treaty, the world still has a huge and deadly arsenal of nuclear warheads that could readily devastate civilization should they ever be used. It's no accident that the initials of the term to describe the standoff of nuclear powers, mutually assured destruction, spell the word they do.

Yet for all the devastating might of current nuclear arsenals, it's easy to forget that we have not gone as far as we could have. In the 1950s, something even more terrifying was predicted. There was every expectation that bombs would be constructed that would not just take out a major city like the thermonuclear weapons of the time, but that could wipe out all life on Earth: a true no-holds-barred doomsday weapon. This hypothetical weapon was sometimes called the C-bomb.

Although the idea had been discussed among the atomic cognoscenti before a single nuclear weapon had been dropped in anger, the doomsday weapon's public unveiling was on an NBC radio show on February 26, 1950. Remarkably popular in an age where intellectual content was still considered appropriate for mass audience broadcasts, this was the *University of Chicago Round Table*, where leading figures would debate a topic of the day.

The subject for this particular discussion was nuclear weapons, and Hans Bethe, one of the lead scientists on the Manhattan Project, warned listeners that one of the worst aspects of the hydrogen bomb was the radiation it could produce. It could, he said, pump radioactive carbon 14 into the atmosphere that would remain a hazard for five thousand years. With enough thermonuclear weapons, the radioactive blanket produced could make life on the planet impossible. But Bethe was to be trumped by Hungarian-born American scientist Leo Szilard, the man who had conceived of the chain reaction, and who made major contributions to the theory behind the atomic weapon. He warned that everything that had been envisaged to date—including hydrogen bombs—could soon be outclassed.

A conventional nuclear weapon caused most of its destruction by its explosion, shock wave, and heat. But as Bethe had described, there was also a more insidious devastation from the radioactive fallout—the longer-term impact of the radioactive residue, spread into the atmosphere by the explosion, causing radiation sickness and death. In the discussion, Szilard imagined a nuclear device where this radioactive side effect would be magnified by encasing a conventional bomb in a jacket of a material that readily absorbed radiation.

This jacket would be vaporized by the explosion, while at the same time the material it was constructed from would be transformed into a highly radioactive form. As the explosion drove the remains of the jacket high into the atmosphere it would spread fine radioactive dust over hundreds or even thousands of miles. With the right material—Szilard suggested the element cobalt would be ideal for the job—this could wipe out whole continents, or even end the life of the whole world.

The model for the way such a bomb would spread dust around the world by first pumping it into the high atmosphere was not any of the bomb tests that had happened in the years leading up to Szilard's remarks on the broadcast. Instead, it was an event that shook the world—literally—on August 26, 1883. It was then that the volcanic island Krakatoa most famously erupted.

Portrayed in the wonderfully inaccurate movie *Krakatoa, East of Java* (Krakatoa is, in fact, west of Java), the volcanic island known locally as Krakatau had seen a number of earlier eruptions, with events recorded as early as AD 416, but the 1883 eruption was massive, producing an explosive power equivalent to a two-hundred-megaton bomb. The significance of this eruption for the bomb makers was not the impact of that explosion, though. It was its ability to spread material around the world.

Around twenty cubic kilometers (four and three-quarter cubic miles) of ash and rock were spewed out by the explosion. Just think of that for a moment—a cube of material nearly five miles on each side. This ash was thrown up eighty kilometers (fifty miles) into the atmosphere and traveled around the Earth, following the eruption's shock wave, which was measured passing around the globe a total of seven times. The result of the dark ash suspended in the atmosphere getting in the way of sunlight was to reduce global temperatures by around one degree Celsius (two degrees Fahrenheit) and to disrupt weather patterns for several years. The ash was detected everywhere around the world. Now substitute for that ash the deadly radioactive fallout of a cobalt bomb and you have Szilard's doomsday weapon.

With a half-life of around five years, the isotope cobalt 60 would have plenty of time to spread around the world, pumping out deadly gamma rays. The impact would be inescapable—there would be nowhere to hide. Such a bomb (or more likely a series of bombs) would need to be much larger than anything that had ever been made. It couldn't be dropped from an aircraft. But it was quite conceivable for such a bomb to take the form of a ship—and the nature of the devastation was such that it would not have to be set off at a particular location.

When Szilard dreamed up this hypothetical weapon, he was aware that there was an obvious question to be answered. Why would anyone other than a madman want such a weapon? Who would want to wipe out all life on Earth? The answer, he suggested, was someone looking for the ultimate in deterrence. If a country was threatened by an enemy with attack, particularly an attack with nuclear weapons, that country could say, "Stop, come no further. If you do, we will destroy the world." As the chairman of the University of Chicago panel pointed out on that radio show, the terrible reality of this kind of weapon is that it would be easier to destroy all life with it than it would be to use the technology to attack an enemy in a controlled way.

While Szilard wasn't sure any existing power would consciously wipe out life from the Earth, he did believe that both the United States and the Soviet Union would be prepared to make such a threat and to construct the means to make the threat meaningful—and if they ever reached a standoff on the use of the doomsday weapon, it was not impossible to imagine them carrying out the threat. Others pointed out that it was quite feasible that a man like Hitler, had he controlled such a weapon, might well have been prepared to use it at the end, when it was obvious that he had lost the war. It does not seem at all fanciful that, unable to control the world, he would have tried to destroy all life on Earth.

Although the idea of a cobalt bomb immediately took on a terrible reality in the mind of the general public, it was only a piece of speculation on Szilard's part when the broadcast was made in 1950. Even so it would reinforce the fear that had already been generated by the dropping of the atomic bomb and that was stoked up by the existence of the hydrogen bomb. This feeling is typified in Tom Lehrer's darkly humorous song "We Will All Go Together When We Go."

The practicality of Szilard's idea was to be given support by other scientists when they worked through the numbers. Admittedly, a doomsday cobalt attack would require something immense—much larger than any atomic weapon that had ever been constructed. It would require thousands of tons of cobalt alone. Yet there was no theoretical limit to the size of a hydrogen bomb—and there was nothing to stop a cobalt doomsday device being made up of a series of ships or land-based sites enabling any size of bomb to be made.

Even though such a bomb was never built, the idea of using radiation as a deadly carpet, rather like the salt that is used in the Bible to seed fields to stop anything from growing, was already in the minds of the public and the military alike. In 1950, in the first year of the Korean War, General Douglas MacArthur proposed following up a defeat of Communist Chinese troops by using radioactive cobalt 60 to produce a five-mile-wide no-go zone between Korea and China making it impossible to pass from one country to the other—the ultimate border control.

MacArthur believed that he had the Communists "in the palm of his hand" and would have been able to crush them by using this tactic had it not been for a combination of harassment and interference from the government in Washington, and the "perfidy" of the British, who MacArthur believed had informed the Chinese of his intentions after being briefed by Washington. Whatever the reasoning, thankfully, the plan to sow a cobalt barrier was never undertaken.

During the cold war, our biggest fear was the outbreak of nuclear war. It was thought that it would bring the end of civilization as we knew it, with only pockets of human survivors expected in the United States, the Soviet Union, and Europe. As far as has been revealed, the closest this came to occurring took place in October 1962, during the Cuban missile crisis. After a spy plane noted a new military site on Cuba, it soon became clear that the Soviets were installing missiles on the island that would be capable of reaching key U.S. cities in minutes.

For the next two weeks, the world teetered on the brink of nuclear war. President John Kennedy described the situation in an address to the nation as "standing before the abyss of destruction." The United States set up a naval blockade of Cuba. Some of the President's military advisers were recommending an immediate preemptive nuclear strike. If it were left to the Russians to attack first, it would be too late.

By the second week of the crisis, the United States was on DEFCON 2, the last state of alert before out-and-out war. Extra bombers took to the skies. In a situation that must have seemed like a living nightmare, the military command prepared to send out the messages that would precipitate nuclear Armageddon. In the end, on October 28, Soviet premier Nikita Khrushchev took a step back from disaster and announced that the missiles would be withdrawn. (The next year, the United States removed some missiles from Turkey that were equally close to the Soviet Union, though this was far less publicized.)

Although nothing came as close as this again, it is sobering to note that the Soviets also set up a doomsday system in the 1970s, designed to cope with the event of an overwhelming attack on the USSR. The idea, much like the hypothetical cobalt bomb doomsday device, was that if the Soviet Union were attacked, retaliation would be so swift and complete that there would be nothing left to fight over. This retaliation would happen even if the conventional Soviet command and control structure was wiped out.

This Perimetr system used a network of computers to assess the situation during a nuclear attack. Should communication with the Kremlin be lost, the system was capable of autonomously issuing the orders to retaliate massively. There was no requirement for human intervention, beyond a confirmation from a relatively junior level. Frighteningly, as far as we are aware, this system, dependent as it is on ancient 1970s computer technology, is still live, still capable of delivering a fatal counterblow to wipe out the Western world.

Despite the end of the cold war, nuclear warfare remains a danger, particularly in tense regions like the India/Pakistan border, and with concerns about development of nuclear weapons in countries like Iran and North Korea, we now also face a more subtle nuclear threat: nuclear terrorism.

The idea that terrorists could bring about nuclear destruction is a horrifying one. There is little doubt that those responsible for the devastation of September 11 would go ahead with a nuclear attack if they were capable of doing so. There seem to be three possibilities for this: building a true nuclear bomb from scratch, obtaining an existing nuclear bomb on the black market, or producing a so-called dirty bomb.

The first requirement to build your own bomb is to get hold of the materials. Occasionally you will see a scare in the press that this might be achieved by the unlikely mechanism of rounding up smoke detectors from stores across the country. Most smoke detectors contain a source of the element americium. Element 95 in the periodic table, americium sits in the detector, beaming out radiation as it slowly transforms to neptunium with a half-life of 432 years. The alpha particles radiating from the americium source (it's a better alpha source than radium) pass through a small compartment where they ionize the air, allowing a tiny electrical current to cross the chamber. If smoke particles get in the chamber, they absorb the alpha particles before they can create ions, stopping the current flowing and setting off the alarm.

It's certainly true that americium can be used to produce a nuclear weapon. Assemble enough of that americium 241 and it will go critical. But before any terrorist groups try to corner the market in smoke detectors it's worth pointing out that it would take around 180 billion of them to have sufficient americium 241 assembled to produce a nuclear device. And even then it wouldn't be enough to put the detectors together in the same place; you would have to painstakingly extract each of those 180 billion specks of the element and mold them together, an effort that would take thousands of years.

At first sight, a terrorist group building a nuclear weapon from scratch seems fairly unlikely even with more conventional radioactive materials. The resources required to build a nuclear weapon have traditionally been huge. Not only do we have the example of the amount of effort that went into the Manhattan Project during the Second World War; there is also the limited success so far from whole countries that have been trying and failing to join the nuclear powers. What hope could a terrorist group have, where an entire country has failed?

However, terrorists could have some advantages over a legitimate state, in that they could obtain fissile material—most likely plutonium, which is produced by the nuclear industries of several countries—illegally, by theft, rather than attempting to purchase it openly or manufacture it. The construction of a traditional bomb itself requires sophisticated engineering and explosives expertise to get the explosive "lenses" that concentrate the impact correctly if plutonium is used—but it is conceivable that a well-funded terrorist group with access to the right expertise could manufacture a crude weapon.

Although plutonium is probably easier to get hold of illegally, the technical problems of making a bomb operate with it are significantly greater than with enriched uranium. Here, to quote nuclear physicist Luis Alvarez, "if highly enriched uranium is at hand it's a trivial job to set off a nuclear explosion. . . . There would be a good chance of setting off a high-yield explosion simply by dropping one half of the material on the other . . . A high school child could make a bomb in short order."

While Alvarez was probably exaggerating a little for effect, a bomb based on enriched uranium could easily be assembled in a garage laboratory. Of course this still leaves a terrorist group with the need to get its hands on enriched uranium. As the Manhattan Project and twenty-first-century examples like the Iranian nuclear program have proved, producing this is nontrivial. But enriched uranium exists in a number of countries around the world, some of them more open to black market deals than others—and its very existence makes it a potential target for theft.

However, the one disadvantage the terrorists have is that enriched uranium, unlike plutonium, which is a natural by-product of electricity-generating nuclear reactors, tends to be in the hands of the military. Uranium enrichment is a tricky, expensive, high-tech process, and there is no reason to have the substance unless you are building a weapon. The confinement of enriched uranium to military establishments does mean it is likely to be more securely stored than plutonium. But there is still the potential for terrorist access where military personnel are susceptible to bribes or threats.

On at least two occasions, quantities of around three kilograms of enriched uranium have been seized from individuals from the former Communist bloc who were attempting to smuggle the materials into the West. Remember that there is no other use for this material than for making a bomb, though it would require around sixty kilograms of highly enriched uranium to make a weapon. We have no idea how many such attempts to smuggle the material have got through undetected.

It seems to some analysts more likely that an existing bomb, from the arsenal of the former Soviet Union or from an area with complex politics like Pakistan, could fall into the hands of terrorists. This is a frightening possibility, though there are some safeguards. Most existing weapons, both U.S. and Soviet, have sophisticated mechanisms to avoid their being used by anyone other than their owners. It's not impossible, but the probability is relatively low. Even so, the chance that terrorists could let off such a nuclear device is one that the U.S. government takes seriously.

In July 2009, the Committee on the Medical Preparedness for a Terrorist Nuclear Event, a group set up by the Institute of Medicine at the government's request, held a workshop that produced a number of recommendations for coping with the impact of a nuclear blast. Although they are eerily reminiscent of the civil defense instructions from the cold war (put tape on your windows, hide under a table), there are serious suggestions here for coping with a nuclear attack.

The point of this work is to make the public more aware of the nuclear threat and to be more prepared. If a nuclear device explodes without warning, there will be little time to issue instructions. It's argued that the population, particularly in target cities like Washington and New York, needs to be aware of what actions they should take in the event of an attack.

The report from the workshop describes the impact of a ten-kiloton nuclear device—the scale of device that is most likely to be assembled by terrorists. Almost everyone within a one-kilometer radius of the blast would die—but outside that area there are possibilities for taking defensive action. Depending on which the way the wind is blowing, the fallout from the blast, a rain of highly radioactive rubble and ash, could plume out for miles from the center of the explosion. It's up to the National Atmospheric Release Advisory Center at the Lawrence Livermore National Laboratory in its California base to predict how the fallout will spread.

It's impractical to outrun this cloud even in a car, so the best advice seems to be to stay indoors with windows closed, where the exposure will be reduced by the building, and put as much masonry and numbers of rooms as possible between the people and the fallout. The best place to be will usually be a basement, followed by the central core of large buildings. According to the committee, just moving to such a protected spot could reduce the risk of immediate death by between a hundred and a thousand times.

However, a dirty bomb seems a more likely approach for terrorists than either building a true nuclear bomb from scratch or getting hold of an existing nuclear weapon. A dirty bomb uses conventional explosives, but instead of surrounding the charge with a payload of shrapnel from nails or similar chunks of metal, the bomb is encased in radioactive material, producing a fallout effect on the area where it is used.

This would deliver a much lower level of radioactivity than the hypothetical cobalt bomb, or even the fallout from conventional nuclear weapons, but would still cause fear and disruption over a wide area. The big advantage for those planning to use such bombs is the ready availability of radioactive material if they aren't fussy about the type they use. No longer is the terrorist limited to fissile elements—the dirty bomb could contain a cocktail from many sources.

Radioactive materials are employed widely in medicine, in agriculture, and in industry for applications from radiotherapy to prospecting for oil. There are a good number of materials that have a long enough half-life to stay around and cause problems for months or longer and that pump out the high-energy gamma rays that cause cell damage and hence radiation sickness. As well as cobalt, the likes of strontium 90 and cesium 137 could all be added to the radiation-producing jacket of a dirty bomb.

There is no doubt that a dirty bomb would cause a terrible incident, and would produce deaths and injuries, but it's probable that the biggest problem it would cause would be fear and disruption. Unlike the fallout from a thermonuclear weapon, a dirty bomb would spread a relatively small amount of radioactive material, and without the devastation of the nuclear explosion, many people would be able to get away before the radiation had a chance to have a serious impact on their health. With a reasonable level of radioactive content it would probably take weeks or months of exposure to provide any significant risk.

However, we can't ignore dirty bombs as a threat. They would spark a huge amount of fear, and could result in devastating financial costs as areas of a city are forced out of use for many months while they are cleaned up, even though the direct threat of the radiation to citizens is relatively low. To give an idea, the level of radiation envisaged from a typical dirty bomb is comparable to the difference in natural radiation levels between Denver and New York. There *is* a relative health risk because of these natural levels. A few more people will get cancer in Denver as a result of it. But it's not something we think of in terms mass destruction.

The U.S. government has taken the problems of nuclear terrorism seriously. Original concerns were mostly that the Soviets would smuggle atomic weapons into the United States or that there could be an accident with an atomic weapon, but in May 1974 the government received a first attempt at extortion based on the use of a nuclear device. A letter was sent to the FBI threatening to explode a nuclear bomb in the Boston area if $200,000 was not paid.

The first of many bluffing attempts to extract money, this helped trigger the government to respond; instead of occasionally using experts from the nuclear labs and the Atomic Energy Commission to assess these kinds of threats, the government set up an ad hoc unit that would be drawn when needed to form what was initially called the Nuclear Emergency Search Team, known as NEST.

Later renamed the Nuclear Emergency Support Team, to emphasize a role that was not just about searching for nuclear weapons but also about rendering them harmless should they be discovered, the team continues to be formed at short notice to the present day—for example, being deployed on a precautionary basis at the Beijing Olympics in 2008.

On the whole NEST activities have, thankfully, been drills or in response to false alarms. Apart from advising on appropriate checks and security, the team is often called in to verify the credibility of extortion attempts, many of which have followed the 1974 Boston incident. Usually the extortionists ask for much more cash, but (as yet) the schemes have never involved actual nuclear materials. This emphasizes once again the relatively low chance of the use of nuclear materials by terrorists (or extortionists)—but doesn't make the vigilance any less necessary.

NEST's role is not made easier by the need to keep what is happening out of the public's awareness, to avoid panic, which could easily cause significantly more casualties than (for example) a dirty bomb. Although originally working on a make-do basis, NEST operatives now have access to a wide range of detectors, built into everything from attaché cases to vans, and airborne detectors in helicopters and planes. All these devices are designed to pick up the almost inevitable stray radiation that is emitted in different forms from a nuclear device.

These detection devices are accompanied by bomb-disposal

equipment, including conventional tools, be they simple wire cutters or supercold liquids to disable detonators, and special apparatuses, such as foam generators producing a material specifically designed to be used to cover a nuclear device and absorb as much of the stray radiation as possible. Proactive presence, exercises, and extortion attempts aside, the only real action the NEST teams have seen is when nuclear weapons or satellites containing nuclear power sources have been involved in a crash or landed in an unexpected site.

As well as the direct activity of NEST, there is also a significant amount of passive detection in place, particular since 9/11. The Department of Homeland Security has been busy installing radiation detectors at ports, airports, and border crossings in the years following the attack on the Twin Towers. As of 2006, around one-third of all shipping containers, and well over three-quarters of the land transport that crosses the Canadian and Mexican borders, were being scanned—and these percentages continued to rise, with all cargo containers crossing the southern border passing through detectors by 2007. Several large cities also have detectors on boundaries and on airborne and van-mounted equipment. This is a threat that is being taken very seriously.

Of course, nuclear weapons aren't the only nuclear threat to our safety. There are also nuclear power plants. Three Mile Island and Chernobyl have demonstrated that things can go wrong at a nuclear power plant. What would happen if we had a real disaster at an American power plant, comparable to or worse than Chernobyl? It's hard to forget that nuclear reactors were first built with the sole intention of producing materials for atomic bombs.

Bearing in mind how much bigger a reactor is than a bomb,

what would happen if one exploded? The possibility is frightening. Yet at the time of writing, nuclear power is being viewed more positively by governments than it has been for many years, because it is (in terms of greenhouse gas emissions) a relatively green way of generating power.

The good news is that a reactor is not going to explode like a bomb. Remember that to make a bomb, the Manhattan Project had to enrich uranium, concentrate the rare uranium 235, or to make plutonium. Reactors run on the much more common uranium 238, which isn't capable of producing the unstable chain reaction needed for a nuclear explosion, and though plutonium can be produced as a by-product, this is removed long before there is enough to cause a threat.

That uranium 238 is not going to run away and explode, because it can't start a chain reaction unless the neutrons emitted by it are slowed down. It needs a moderator, a material like carbon or water that will slow the neutrons enough to give them the time needed to cause fission with uranium 238. Although it is possible to have a nuclear accident, it isn't possible for the whole thing to explode like a nuclear bomb—if the reactor started to run away, producing the kind of fast neutrons necessary for a bomb, the reaction would automatically cease.

That's not to say that nuclear reactors can't explode. But the explosion is a conventional one, caused by too much heat in a confined space, not a nuclear explosion. We can gauge the impact of failures of control in nuclear power stations from two well-publicized incidents: Three Mile Island and Chernobyl.

The accident at the Three Mile Island reactor, sited on an island in the Susquehanna River near Harrisburg, Pennsylvania, was by

far the less significant of the two, even though its name is synonymous with dangerous nuclear accidents. The water-cooling system
at the power plant failed, and there was a series of errors in the
deployment of fallback systems. The containment vessel, the steel
bottle that keeps all the nasty material in the reactor, was *not*
breached—the reactor did not melt down—but there was some
leakage of coolant that carried radioactive material, and gas was
intentionally vented from the containment vessel to keep the pressure under control, again sending radioactive material out of the
plant. The result was a low level of exposure that could have resulted in perhaps one death.

The Three Mile Island incident was blown up out of all proportion, so that it is still thought by many to have had a much greater
impact than it actually did, for two reasons. First, the problem was
picked up by those with an interest in shutting down nuclear
plants, and used as a rallying cry about the deadly dangers involved. No one is saying that the accident wasn't a bad thing. But
there are many more industrial accidents each year causing greater
death and destruction—it's just that the other industries don't
have the same strength of feeling against them.

The other problem arose from the indiscriminate use of Geiger counters. Worried about the radioactive leaks from the plant,
amateurs measured radiation levels around the area and were
horrified to discover that the readings were as much as 30 percent above the national average. This sounded scary. This level
was high enough to cause up to sixty times as many deaths as the
official impact of the leak. Clearly, it was thought, there was a
cover-up. Only there wasn't. This was just the natural level of
radioactivity in the area, caused by the radioactive radon gas that

is naturally produced from such neighborhoods built on rocks like granite.

It puts the impact of the Three Mile Island accident into perspective that the natural radiation danger from just living in that area, a danger that was present before the power station was built and would continue even if it were totally removed, was sixty times higher than the risk from the accident itself—and some areas have even higher natural background radiation levels.

Three Mile Island, then, was not as bad as it is often portrayed. But no one can suggest that Chernobyl was a minor incident. The worst nuclear power station accident the world has seen, Chernobyl in Ukraine really did suffer an explosion leading to a breach of the containment vessel, the nightmare scenario for anyone dealing with nuclear safety.

On the night of April 25, 1986, engineers at the Chernobyl nuclear plant carried out a planned test of the emergency core cooling system, the central system that ensures that the operational part of the nuclear reactor does not overheat to such an extent that it disrupts the massive vessel that contains it.

As the test commenced, an operator made an error, bringing the reactor to a nearly shut-down condition. To have enough power to continue the test, engineers had to manually override the safety system to be able to withdraw the control rods that moderate the nuclear reaction. As the engineers repeatedly canceled safety warnings, temperature soared, a massive buildup of steam blasted the top off the reactor, and the Chernobyl disaster began to unfold. (Note once again, this was not a nuclear explosion like that from a nuclear bomb, it was steam pressure that blew the containment vessel open.)

That this could happen is partly attributable to the bad design of that plant. Most nuclear reactors are designed to be self-regulating systems. As they heat up, neutron activity increases, and the result is loss of the chain reaction, because the neutrons are whizzing around too fast to cause a chain reaction in uranium 238. But the Chernobyl design *increased* the reaction rate as temperature increased, providing a positive feedback loop that, with control systems inactive, meant it was impossible to overcome the problem. Worse still, the reactor vessel was not inside the type of concrete containment building used in U.S. designs. It blew open its metal container to send radioactive materials spewing into the atmosphere. The smoke billowing from the ensuing fire contained massive amounts of radioactive particles.

The result was a spate of deaths, over the few days following the explosion, among those exposed to deadly amounts of radiation as they tried to control the fire, followed by a long-term death toll that could have reached as much as four thousand (this number has subsequently been queried) as cancer rates in nearby areas shot up, particularly in neighboring Belarus in the direction the wind was blowing.

The commercial impact of the accident was huge too. Hundreds of miles away in the United Kingdom, sheep flocks had to be destroyed because the radioactive levels in the grass they were eating had temporarily increased (though no one in the United Kingdom is thought to have suffered enough of an increased exposure from the Chernobyl fallout to be at risk). Areas of the countryside around Chernobyl were evacuated and are still considered unsafe.

What should be repeated, though, because there is often still misunderstanding about out-of-control reactors, is that there was no runaway chain reaction causing a nuclear explosion. The chairman of the Senate Intelligence Committee made a public statement that the Soviets were lying about what was happening at Chernobyl because they said that the loss of coolant meant the chain reaction had stopped. They weren't—it had. The coolant has a secondary effect of slowing down the neutrons emitted, and without slow neutrons there is no chain reaction. It stops dead. There is still plenty of radioactivity—and much too much heat—but no potential for a runaway chain reaction leading to nuclear explosion.

This was, by any means of measurement, a terrible accident. Yet there have been even worse industrial accidents, killing more people, causing more devastation. And the impact of Chernobyl on the environment around it has proved less dramatic than was first thought. There is still contamination, but studies of the flora and fauna around Chernobyl show that they are surprisingly normal. Those used to postapocalyptic landscapes from fiction may also be surprised that giant rats and cockroaches haven't taken over. In fact, rodents have a bigger tendency to die off than the larger animals, and cockroaches apparently don't deserve their reputation for being good at resisting radiation either.

Rather than being a blasted wasteland, much of the abandoned territory around Chernobyl teems with wildlife—even bear, beaver, lynx, and bison which rarely survive in Europe. Apart from some distortion in trees, particularly pine, whose sticky coatings tended to hang on to radioactive dust, there is little obvious mutation. There are no monster creatures, or fish with three eyes as those familiar with the nuclear plant in *The Simpsons* might expect. In fact, mutation among surviving animals seems rare—apparently the most common effect of exposure to nuclear radiation is for animals to die, and those that do survive in a damaged form are less attractive for breeding, so mutants have not transformed the biological landscape.

We mustn't underplay the significance of Chernobyl. Due to a combination of poor design, bad systems and maintenance, and incompetence on the part of the operators, the accident caused terrible damage and many avoidable deaths. Yet in apocalyptic terms, this was no Armageddon. The impact was smaller than an industrial accident like Bhopal, and vastly smaller than the destruction caused by many of the wars that have taken place over the last hundred years.

There remains the concept of the China Syndrome. With reactors unable to provide a true nuclear explosion, this is usually put forward as the ultimate disaster scenario for a nuclear power plant, as portrayed in the movie of the same name. It's a truly frightening idea. The dramatic name implies a runaway meltdown that doesn't just make the reactor building collapse. Instead, the overheating reactor eats through the land beneath it and goes on to melt its way through the Earth's core, all the way though the planet until it pops out in China.

If anything even vaguely like this were to happen it would be truly catastrophic. You don't even have to get all the way through. Unlike Jules Verne's friendly picture of a network of caverns under the ground in *Journey to the Center of the Earth,* there is intensely hot liquid rock down there, plus natural nuclear reactions that maintain the heat, and immensely high pressure. If a sustained meltdown plunged deep enough into the Earth it could cause the supervolcano to end all supervolcanoes.

Thankfully, though, dramatic though the image is, it just couldn't ever happen. Leaving aside the lack of nuclear reactors on the side of the planet opposite China (despite popular public opinion, China is not on the opposite side of the globe from the United States; it's opposite parts of South America and a lot of ocean), a nuclear reactor could not melt far into the Earth's crust. It wouldn't stay a concentrated, superhot blob, but would spread out and cool on contact with the surrounding ground, its temperature rapidly dropping back to normal levels.

As we've seen, the poor design of Chernobyl's reactor was part of the reason for its failure. All Western nuclear power stations were already safer before the accident—and the safety of the USSR's stations was enhanced after the disaster. There is, however, a way to make generating nuclear power far safer still. There is a totally different design of reactor called a pebble-bed reactor, where the uranium sits in a bed of pebbles made of a special type of graphite (carbon).

The key safety element of the pebble-bed reactor is that it is inherently not susceptible to the usual cause of reactor failure. As we've already seen, if a reactor loses cooling its chain reaction stops, and the danger comes instead from the way the intense heat produced can set the graphite control rods alight and the resultant fire can melt through the containment vessel. In a pebble-bed reactor, the pebbles are made of pyrolytic graphite, which is so resistant to heat that the temperatures reached would not cause them any damage. There would be no meltdown, no fire to damage the containment vessel. If the reactor ran out of control it would simply heat up, peak, and cool down without ever threatening its environment. What's more, such reactors are also more efficient than the traditional design because they run at a higher temperature.

It is bizarre that as the nuclear states build more nuclear power plants to manage their electricity needs with less global warming, few seem to be seriously considering pebble-bed reactors. It would be a major step away from possible future nuclear disasters.

The other alternative to make nuclear power safer is nuclear fusion. Here the power of the Sun is brought to Earth. Given that fusion is the process that makes the difference between a thermonuclear bomb and a conventional fission bomb, it might seem that fusion power stations would be taking nuclear madness to a new level, but in fact fusion power is both vastly safer and cleaner in terms of nuclear waste than is fission.

The existing fission power stations require a rare ore to be mined to produce uranium, and when they are used, the power stations produce long-lived, highly radioactive by-products (including plutonium) that need to be stored for many thousands of years and kept safe from terrorists. Fusion, by contrast, uses abundant hydrogen as fuel rather than those rapidly depleting stocks of uranium, and produces only low-level radioactive waste (though the fusion vessel will become radioactive over time). What's more, fusion reactions are hard to keep going—they have built-in safety because should anything go wrong, the whole reaction instantly stops.

But that "hard to keep going" is as much a problem as it is an advantage. Despite nearly fifty years of research, we are still without a working fusion reactor that produces more energy than it needs to get it going in the first place. The next-generation device, which is expected to get to that self-sustaining state, is the International Thermonuclear Experimental Reactor (ITER) an internationally funded project to be based in France and expected to be operational by 2016. (At the time of writing, the U.S. government has just, with supreme lack of foresight, cut $160 million from its contribution to ITER, which may well result in the completion date being pushed back.)

Assuming ITER is a success, plans are to have a commercial model available by 2050, though cynics point out that ever since the 1950s it has been said that a commercial fusion reactor is about thirty years away. There is a lot of hope, but the time frames here are such that nuclear fusion could not provide a sizable part of our power needs until somewhere between 2070 and 2100. The promise exists of nuclear power without the possibility of a hugely destructive accident—but it isn't going to be available for a long time.

Some would argue that the next threat to humanity we will explore—climate change—is also a long way from being a meaningful danger. And it's true that the estimates of the risk from climate change have tended to change over the last few years. But every new piece of data we have seems to make the threat more real, and more immediate.

CHAPTER FOUR

CLIMATE CATASTROPHE

||||||||||||||||||||||||||||||||||||||||||||||||||||

> *Since 2001, there has been a torrent of new scientific evidence on*
> *the magnitude, human origins and growing impacts of the*
> *climactic changes that are under way. In overwhelming*
> *proportions, this evidence has been in the direction of showing*
> *faster change [and] more danger.*
> —John P. Holdren, president of the American Association
> for the Advancement of Science, quoted in *The New York Times*,
> February 2, 2007

It's only now that we are beginning to face up to the reality of the impact that our everyday science and technology has had on the environment. You don't need to look at some evil weapon of mass destruction conjured up by a mad scientist—our transport, housing, industry, and consumption in general are having a direct effect on the world we live in. Climate change is already under way. Just a few degrees of global warming would be enough to bring worldwide civilization to the verge of collapse.

This is the most insidious way that science can present a threat to humanity. We have all benefited hugely from the mechanization of civilization. We can achieve actions that once would have seemed incredible—like flying from one side of the planet to the other—with very little personal effort. But we are just beginning to realize the impact that our life-enhancing science and technology is having on the planet.

It would be fair to say that until recently, the threat from climate change was not well understood. In my *Oxford Dictionary of Scientific Quotations,* published in 2005, there is no reference to global warming and only one mention of climate change—and this is unconnected with the threat with which we are now concerned. Just take a look at that quote from the American atmospheric chemist Richard P. Turco, made in 1983, to see how much things have changed.

> Global nuclear war could have a major impact on
> climate—manifested by significant surface darkening over
> many weeks, subfreezing land temperatures persisting
> for up to several months, large perturbations in global
> circulation patterns, and dramatic changes in local
> weather and precipitation rates—a harsh "nuclear winter"
> in any season.

In a world still living under the threat of imminent nuclear war, it seemed that humanity's main impact on the climate could be nuclear winter. The vast amount of smoke and debris from the explosions of atomic bombs would, like Krakatoa's haze of dust writ large, act as a sunshade in the atmosphere long enough to seriously chill the planet, perhaps even bringing us to the apocalypse of a world undergoing a massive ice age that would wipe out life on the planet in all but a narrow equatorial band.

As the threat of atomic devastation has become less significant, a very different, much more subtle type of climate change has reared its head. This is the kind of change portrayed in Al Gore's movie, *An Inconvenient Truth*. For a long time this has been played down or even dismissed by the press and those with vested interests in ignoring climate change. And it's true that portrayals like *An Inconvenient Truth* have not always helped, because they have tended to be a little careless with the facts in their enthusiasm to get the message across. But we shouldn't take this as meaning that climate change is not happening, nor that its impact on human life will be insignificant.

The obvious sources of opposition to the findings of environmental science are companies benefiting from products and services that threaten the environment. It's not a simple equation that everyone involved in a polluting company is necessarily a bad guy. The executives of these companies are human beings with children—they may well want to help the environment. Yet history has shown that commercial organizations are very good at ignoring the negative aspects of their products until they are forced to take them into account. The past actions of cigarette manufacturers make it clear that companies are prepared to ignore evidence until the last possible moment, and to try to manufacture opinion that supports their business objectives.

Time and time again those who have a motive to suppress the bad news about climate change have ignored evidence and tried to counter expert views. Often such actions are done through third parties—organizations and individuals that present the anti–global warming message in an apparently independent manner, but whose funding can be traced back to energy companies and others businesses that find climate change a commercially irritating concept.

In early 2007, Senator Barbara Boxer, chair of the U.S. Senate's Environment Committee, had a meeting with the head of the Intergovernmental Panel on Climate Change (IPCC), the body set up by the World Meteorological Organization and the UN to provide scientific evidence to governments around the world. At this meeting, Senator Boxer was given an unequivocal message that climate change was real, and there was a very high probability that the burning of fossil fuels was a major contributor to the problem. As she left the meeting, one of Senator Boxer's staff pulled her to one side. She was told that a conservative organization, funded by an oil company, was offering scientists $10,000 to write articles that attacked the IPCC report and the models that had been used to produce its gloomy predictions.

This was organized resistance. Former senator Tim Wirth, onetime Democratic spokesman on the environment, drew a parallel with the tobacco industry at its height. "Both figured, sow enough doubt, call the science uncertain and in dispute. That's had a huge impact on both the public and Congress." The result has been to produce significant doubt and confusion in the public's mind. The message the public has received from much of the media is that scientists and science are divided on whether or not human-caused climate change exists.

Those who feel that the whole idea of a covert alliance attempting to dismiss global warming smacks too much of conspiracy theory might be shocked to learn that as long ago as 1998, a group including representatives of well-known organizations that argue against global warming met with representatives of oil company Exxon at the hardly unbiased American Petroleum Institute to discuss a campaign to train twenty scientists to become media representatives for their viewpoint. (This campaign was quietly dropped when memos from the meeting were leaked.)

ExxonMobil seems finally to be losing its will to keep fighting the anti–global warming fight. After being rapped on the knuckles by the U.S. Senate for spending over $19 million on anti–global warming organizations producing, as one senator put it, "very questionable data," the company has publicly announced that it accepts the risks posed by climate change. These could be weasel words, and there are still senior Republican politicians who have been unable to shake off the anti–global warming view—but the United States seems to be finally turning the corner on climate change.

Part of the problem with understanding the likely consequences of continuing to abuse the environment on the scale that we do is that the threats don't always sound particularly scary. In worst-case scenarios there is talk of temperatures rising as much as five degrees Celsius (nine degrees Fahrenheit) by the end of the century. This doesn't sound particularly scary. Those of us who live at chillier latitudes may even think a few degrees warmer wouldn't be a bad thing. But there's a wealth of unpleasant detail hiding beneath those small numbers.

First, the numbers are averages. A rise of a few degrees in the average temperature can mean peak values that soar far above current levels. Averages are useful to get a broad picture of what is happening but can be extremely misleading when we try to understand what we experience. Just consider the difficulty of making deductions from averages. The average person has fewer than two legs (because some are missing a limb, the average is below two). Does this mean that shoe stores should stop selling shoes in pairs?

It would be a ridiculous act—but that's what would happen if you *only* considered averages. We don't live the averages; we live through the peaks and troughs—however extreme they may be.

And then there's the wider impact of climate change. It's not just about the temperatures rising. Accompanying warming of this scale are impacts like droughts, the sort of wildfires that have swept California in the past occurring much more frequently, and sea-level rises from melting ice that could see whole low-lying swaths of real estate left useless. Think of what happened in New Orleans after Hurricane Katrina struck in 2005, repeated on most of the low-lying coastal areas around the world.

We are reluctant to do anything about climate change, because preventing it is expensive, and it requires us to suffer financial pain now to deal with a future problem that can't be exactly quantified—a difficulty that became doubly problematic as the world sank into recession in 2008 and 2009 and most governments decided that getting the economy going again was more important than thinking about the planet. We saw, for example, programs to encourage us to go out and buy more cars. This was great for jobs, but not so great for the planet.

It's shocking how long it has taken for there to be widespread acceptance that climate change is really happening. It shouldn't be news. The U.S. National Academy of Sciences made its first study of global warming back in 1978. Although widespread acceptance that there is a serious problem took time to develop, the impact of climate change has now been studied for a good number of years—and the vast majority of scientists accept that this change is strongly influenced by human activity.

The UN added its support in the form of the 2007 report from the Intergovernmental Panel on Climate Change, stating that global warming is a fact, and that most of the rise in temperature since 1950 is most likely (with a better than 90 percent confidence) to have been caused by human intervention. "February 2 [of 2007] will be remembered as the date when uncertainty was removed as to whether humans had anything to do with climate change on this planet. The evidence is on the table," said Achim Steiner, executive director of the UN Environment Programme.

Even the few scientists who don't accept a man-made component to climate change admit that we are undergoing global warming. According to the IPCC, the world can look forward to centuries of climbing temperatures, rising seas, and disrupted weather. The ten warmest years on record have all occurred since 1990, and most of those were in the last decade. All the evidence is that the world is warmer now than at any time in the past two millennia—if current trends continue, by the end of the century it will be the hottest it has been in 2 million years.

There is a lot of talk about action to prevent climate change—but, realistically, this is not likely to have enough effect. It will almost certainly be a matter of too little, too late. Even if we persuaded the Western world to give up its love affair with the SUV and cheap flights, the economies of China and India are gearing up to rival those of the West. It has been argued that the only way to prevent climate change passing through a tipping point where warming will accelerate beyond our control is to reduce greenhouse gas emissions by 90 percent by 2030. No politicians are suggesting cuts that will achieve anywhere near this level of reduction.

We don't have to reach that tipping point to see climate change accelerating. Already the trends are getting worse. As *New Scientist* magazine said in February 2007, "The [IPCC] authors acknowledge that they were being conservative. There is, though, a fine line between being conservative and being misleading, and on occasion this summary crosses the line. It omits some real risks either because we have not pinned down their full scale or because we do not yet know how likely they are." Every week brings new revelations that global warming will hit us harder and sooner than was previously thought.

Apart from a relatively small impact from the heat of the Earth's core, the world's warmth comes from the Sun. Without the energy of sunlight, the surface of the Earth would be similar to that of a distant planet in the solar system with a temperature hovering below −240 degrees Celsius (−400 degrees Fahrenheit). The Sun's warmth is essential to preserve life—but it is also the Sun that pushes us into global warming. Normally a fair amount of the Sun's energy is reflected back off the Earth out into space. The more of that energy that is absorbed by the atmosphere and the planet, rather than reflected, the more Earth's temperatures will rise.

The greenhouse effect, which we've heard so much about, modifies the amount of the Sun's energy that escapes back through the atmosphere. Again, like the Sun, this isn't a bad thing in itself. If there were no greenhouse effect, the Earth would be an unpleasantly chilly place, with average temperatures of −18 degrees Celsius (zero Fahrenheit), around 33 (60) degrees colder than it actually is. But living in a gaseous greenhouse can be just as troublesome as not having its protection.

The greenhouse effect is caused by water vapor and gases like carbon dioxide and methane in the atmosphere. Most of the incoming sunlight powers straight through, but when the energy heads back into space as infrared radiation, some of it is absorbed by the gas molecules in the atmosphere. Almost immediately the molecules release the energy again. A portion continues off to space, but the rest returns to Earth, further warming the surface.

We only have to look into the sky at dusk or dawn when the planet Venus is in sight to see the result of a truly out-of-control greenhouse effect. Venus is swathed in so much carbon dioxide (around 97 percent of its atmosphere) that relatively little energy ever gets out. Admittedly our sister planet is closer to the Sun than is the Earth, but it's this ultrapowerful greenhouse effect that results in average surface temperatures of 480 degrees Celsius (900 degrees Fahrenheit)—hot enough for lead to run liquid—and maximum temperatures of around 600 degrees Celsius (1,100 degrees Fahrenheit) making it the hottest planet in the solar system.

No one is suggesting that the Earth's atmosphere is heading for Venus-like saturation of greenhouse gases, but there is no doubt that the concentration of carbon dioxide, methane, and other gases that act as a thermal blanket is going up. Each year we pour around 26 billion tons of carbon dioxide (CO_2) into the atmosphere. Around a quarter of the CO_2 we produce is absorbed by the sea (though this process seems to be slowing down as the oceans become more acidic), and about a quarter by the land (much of it eaten up by vegetation), but the rest is added to that greenhouse gas layer.

Looking back over time—this is possible thanks to analysis of bubbles trapped in ancient ice cores from Antarctica and Greenland, where the further down we drill, the further back we look in time—the carbon dioxide level was roughly stable for around eight hundred years before the start of the Industrial Revolution. Since then it has been rising, and even the rate at which it rises is on the increase—not only is the level of CO_2 in the atmosphere growing; the growth is accelerating.

In preindustrial times, the amount of carbon dioxide in the atmosphere was around 280 ppm (parts per million). By 2005 it had reached 380 ppm, higher than it has been at any time in the last 420,000 years. It's thought that the last time there was a consistent comparable level was 3.5 million years ago in the warm period in the middle of the Pliocene epoch, well before the emergence of Homo sapiens, and it's likely that levels haven't been much higher since the Eocene epoch, 50 million years ago. The Intergovernmental Panel on Climate Change predicts that if we don't change the amount of CO_2 we generate, levels could be as high as 650 to 1,000 ppm by the end of the century. The Goddard Institute for Space Studies (GISS) model, one of the best computer simulations of the Earth's climate, which reflects the impact of these changes on water patterns, predicts that most of the continental United States will regularly suffer severe droughts well before then.

Current predictions are that by the end of the century, the tropics will live through droughts thirteen times as often as they do now. Drought is already on the increase. A 2005 report from the U.S. National Center for Atmospheric Research notes that the percentage of land areas undergoing serious drought had doubled since the 1970s. Southwestern Australia, for instance, is facing a steady reduction in rainfall, leading to both potential drought and increased chances of bushfires.

As drought conditions spread, availability of water becomes restricted. Significant decreases in water output from rivers and aquifers are likely in Australia, most of South America and Europe, India, Africa, and the Middle East. Across the world, drought will be dramatic. The 2007 report of the UN Intergovernmental Panel on Climate Change predicted that by the last quarter of the century between 1.1 billion and 3.2 billion people will be suffering from water-scarcity problems.

Most historical droughts have been relatively short-term. Caused by statistical blips in the climate rather than marked permanent change, they cause devastation and disaster, but can be recovered from. A long-term drought provides no way out. Where these have happened in the past, civilizations have simply disappeared. After three or four years, the inhabitants of the drought area are faced with a simple choice: evacuation or death. A couple of years later and you have an abandoned region, littered with ghost towns and dead villages. Drought is no minor inconvenience.

At first glance, the whole concept of running low on water is an insane one. Looked at from space, the defining feature of the Earth when compared with the other planets in our solar system is water. Our world is blue with the stuff. In round figures there are 1.4 billion cubic kilometers (a third of a billion cubic miles) of water on the Earth. This is such a huge amount, it's difficult to get your head around. A single cubic mile (think of it, a cube of water, each side a mile long) is around 1 trillion gallons of water.

Divide the amount of water in the world by the number of people and we end up with nearly a tenth of a cubic mile of water each. More precisely, 56 billion gallons for everyone. With a reasonable consumption of 1.3 gallons per person per day, the water in the world would last for 116,219,178 years. And that assumes that we totally use up the water. In practice, much of the water we "consume" soon becomes available again for future use. So where's the water shortage?

Things are, of course, more complicated than this simplified picture suggests. In practice, we don't just get through our 1.3 gallons a day. The typical Western consumer uses between 1,500 and 3,000 gallons. In part this happens directly. Some is used in taking a bath, watering the lawn, flushing the toilet—but by far the biggest part of our consumption, vastly outweighing personal use, is the water taken up by manufacturing the goods and food that we consume. Just producing the meat for one hamburger can use 1,000 gallons, while amazingly, a one-pound can of coffee will eat up 2,500 gallons in its production.

However, even at 3,000 gallons a day, we still should have enough to last us over 57,000 years without even adding back in reusable water. So where is the crisis coming from? Although there is plenty of water, most of it is not easy to access. Some is locked up in ice or underground, but by far the greatest majority—around 97 percent of the water on the planet—is in the oceans.

For countries with a coastline, like the United States, this is not particularly difficult to get to, but it is costly to take seawater and make it drinkable. The fact that nations with coastlines are prepared to spend huge amounts of money on reservoirs to collect a relatively tiny proportion of fresh rainwater, rather than use the vast quantities of sea that border them, emphasizes just how expensive is the desalination process required to turn seawater into drinkable freshwater.

Water shortages, then, come down to a lack of cheap power. If we had unlimited extremely cheap power, there would not be a water shortage. More indirectly, the price of power also limits our access to food. Drought makes food harder to grow, since we must rely more on expensive irrigation; but with sufficient power, irrigation should not be an issue. On the world scale, as climate change bites, limits on power availability make it harder to provide irrigation and to transport food around the world to meet global need.

Even where there is not the immediate threat of drought, the rise in temperature can push previously lush areas into decline. Many areas that are currently tropical forests—the Amazon rain forest has to be the best known example—are predicted to change to savannah, grassland, or even desert as carbon dioxide levels rise and a combination of lack of water and wildfire destroy the woodland. The Amazon, long touted as the lungs of the world, has already become an overall source of carbon dioxide, pumping over 200 million tons of carbon from forest fires into the air—more than is absorbed by the growing forest. If things continue the way they are, the expectation is that the Amazon rain forest will be just a memory by the end of the century.

This change of the environment from carbon sink—a mechanism to eat up carbon dioxide from the air—to carbon source is a feature of not just tropical forests. In 2005, scientists in the United Kingdom reported that soil in England and Wales had switched from being a carbon sink to being a carbon emitter. As average temperatures rise, the bacteria in the soil become more active, giving off more CO_2. Remarkably, in 2005 this was already proving enough of a carbon source to cancel out all the benefits from reductions in emissions that the United Kingdom had made since 1990.

A combination of decrease in rainfall over areas like the Amazon rain forest with increase in temperature is expected to result in a massive die-off. There is a similar expectation that temperate and coniferous forests in Europe and parts of North America will be drastically reduced. The picture isn't uniformly gloomy—there is some expectation of a northern expansion of forest in North America and Asia—but even so, the overall effect is that vegetation that has been soaking up carbon will, in our lifetimes, reverse to being an overall source of carbon, kicking the greenhouse effect into positive feedback. And positive feedback is the worst possible news about climate change.

The best-known example of positive feedback is the howl from a sound system when a microphone is brought too close to the speaker. Tiny ambient sounds are picked up by the microphone, come out of the speaker louder, are collected again by the microphone, and are reamplified, getting louder and louder until they become an ear-piercing screech. One of the most worrying aspects of climate change is that the global climate also features a number of positive-feedback systems, where a change reinforces the cause of the change, making the change happen faster, which reinforces the cause more, and so on.

Positive feedback has often been omitted from predictions. As *New Scientist* put it in February 2007, "The rising tide of concern among researchers about positive feedbacks in the climate system is not reflected in the [IPCC report] summary. . . . One clear need is to get to grips with the feared positive feedbacks."

It's not just the Amazon rain forest and the Australian bush that are tipping into positive feedback, adding to the greenhouse effect. Other forests around the world are being taken out of the carbon-sink equation as temperatures rise. For example, a combination of the increased temperature and the spread of pests is having a devastating effect on some Canadian forests. In one year, British Columbia lost nearly one hundred thousand square kilometers (forty thousand square miles) of pine trees (over half the land area of the state of Washington) to a combination of forest fires and disease. The local government estimates that 80 percent of the area's pines will be gone by 2013.

Wildfires, destroying thousands of hectares of land and properties, are becoming increasingly common. In 1998 fires destroyed 485,000 acres in Florida and 2.2 million acres in Nicaragua. This is happening more and more frequently. More than 600,000 acres were destroyed on the Florida/Georgia border in 2007. Even in previously temperate areas like the United Kingdom, wildfires now pose a threat.

Agriculture will be forced to undergo major changes. Traditional crops of hot countries will take over in previously temperate regions, while areas already growing such high-temperature crops will find it increasingly hard to provide any food. The 2007 IPCC report that forecast huge water shortages also predicted that as the twenty-first century progresses, up to 600 million extra people will go hungry as a direct result of climate change.

If things get too drastic, perhaps our only hope will be a "Noah's ark" of food—the vault being built by the Global Crop Diversity Trust in the permafrost of the Svalbard Archipelago near the North Pole, which will contain 3 million batches of seeds from all current known varieties of crops as a defense against the impact of global catastrophe.

There is an even more insidious effect of global warming that provides another, particularly dramatic, positive-feedback loop in the climate system—the melting of the Siberian permafrost.

In western Siberia lies a huge peat bog, around 900,000 square kilometers (350,000 square miles) in area—the size of Texas and Kansas put together. Peat, the partly decayed remains of ancient moss and vegetation, is a rich source of methane, a gas that contributes twenty-three times as much to the greenhouse effect weight for weight as does carbon dioxide. The methane from the bog is frozen in place by the permafrost—a solid mix of ice and peat that never melts. At least, that never melted until now. That permafrost is liquefying, discharging a huge quantity of methane into the atmosphere. By 2005 it was estimated that the bog was releasing 100,000 tons of methane a day. That has more warming effect than the entire man-made contribution of the United States. And thanks to positive feedback, the more the bog releases methane, the faster it warms up, releasing even more.

The impact of increasing temperatures is even worse for our city dwellers than for the rest of the population, thanks to the urban heat island effect. In a normal environment, summertime temperatures are kept under control by nighttime cooling. Without energy from the Sun hitting the dark side of the Earth, the planet can only lose heat, and where there are clear skies this can happen surprisingly quickly, providing the biting cold nights of the desert. But something goes wrong with this natural cooling process in a city. The sidewalks and canyonlike streets act as storage heaters, absorbing energy that will keep temperatures relatively high at night.

This is the reason that many of the casualties of the European heat wave of 2003 were in cities. It's not a sudden, short snap of heat that is a large-scale killer; it's sustained heat that goes on day after day, and particularly that carries on through the hours of darkness. In the 2003 heat wave, it never got cool enough at night for relief. On August 12, 2003, Paris suffered a nighttime temperature that never went below 25.5 degrees Celsius (78 degrees Fahrenheit), stifling for the majority of city-center households without air-conditioning. Thousands died from the impact of the relentless heat held in place by the city streets. The final European death toll was over thirty-five thousand from the heat and up to fifteen thousand more from the pollution that built up, particularly over cities, in the warm still air.

Europe isn't alone in suffering the impact of sustained heat. Even though air-conditioning is much more widespread in the United States, hundreds died in Chicago in July 1995 when a heat wave of such sustained ferocity hit the city that on two successive nights the thermometer never dropped below 27 and 29 degrees Celsius (80 and 84 degrees Fahrenheit) respectively. To make matters still worse, warm air rises. The temperature difference between the ground floor and the top floor of a building can be enough to make the difference between comfort and trying to sleep in a virtual oven. Older high-rise buildings without air-conditioning but with relatively good air flows are particularly susceptible to roasting inhabitants on upper floors.

The urban heat island effect is real, and because it is factored out of climate change calculations to avoid confusing the impact from greenhouse gases, it means that cities are likely to fare significantly worse than the predictions of temperature rise given by the climate change models. It has been shown that urban heat islands don't contribute particularly to the overall warming of the planet (this can be seen because there is no real difference in global temperature between still days in the city, when the effect arises, and windy days)—but that really doesn't matter to the person in the city apartment. She will still suffer much more than the models predict.

That's just how things are now. Killer heat waves like the one that struck Chicago in 1995 currently might be expected every twenty years or so, but are likely to be annual occurrences by the end of the century according to our best predictions. It is quite likely, according to today's forecasts, that a summer with such temperatures would be average by 2040 and could be typical of the coldest summer of the decade by the 2060s. The one mitigating factor that might benefit some northern areas is the slowing down of the thermohaline circulation, the complex system of ocean currents that transport large amounts of heat from the tropics to northern latitudes. The section of this ocean conveyor system that runs in the surface layers of the Atlantic, stopping the northeastern coastline of the United States and northern Europe from being more like Siberia in temperatures, is the Gulf Stream.

There is some evidence that climate change could produce a reduction of strength in this ocean conveyor, because of the impact of freshwater from melting ice sheets. The collapse of the conveyor was the scenario dramatized in the movie *The Day After Tomorrow*, but this hugely overemphasized both the speed of the change and its impact. Early attempts to model the impact of climate change on the conveyor suggested that it might shut down entirely over this century, but current best estimates predict a decrease in strength of around 25 percent. This will help mitigate the heat impact of climate change in the areas warmed by the Gulf Stream, but will not totally counter it.

As the planet warms up, the delicate balance of coastal life will be devastated. Increasing temperatures inevitably result in sea-level rises. This is not just a case of irritating the coastal wildlife. Many of the world's great cities from New York to London, and major sections of low-lying countries like Bangladesh, are at risk of destruction by relatively slight increases in sea level. In the storm surge of 1998, 65 percent of Bangladesh was inundated. It would not take much of a rise to make this a permanent state.

Climate change has a double impact on sea level. The headline-grabbing cause is the melting of vast tracts of ice, increasing the volume of water as they plunge into the sea, but there is a more direct effect too. As liquids get warmer they expand, and with the huge volume of water in the oceans this has a significant effect. Just a few degrees' increase in temperature is enough to push the sea level up several feet from expansion alone. But the melting ice isn't just featured more often in the news because it looks more dramatic on the TV screen. Though currently expansion is responsible for more rise than melting, the situation with the world's frozen supplies of water is heading for potential catastrophe.

A very visual illustration of the impact of climate change is the way that ice is disappearing from the North Pole in the summer. Not only is this happening on a large scale, but also Arctic ice is melting much faster than was expected only a few years ago. NASA satellites have revealed that between the winters of 2004 and 2005, three quarters of a million square kilometers (280,000 square miles) of ice that was normally permanently frozen melted: this is without historical precedent. The summer low of 2005 had a polar ice cap with 20 percent less area than that of 1978. At least once in the last few years, the North Pole itself has disappeared. This can happen because the Arctic isn't a landmass but a floating sheet of ice.

The good news here is that melting Arctic ice doesn't contribute to an increase in sea levels. Floating ice is already displacing water just as a ship does—if the floating ice melts, the overall water level doesn't rise. But that doesn't mean the disappearing Arctic summer ice is a good thing. Not only is it a disaster for wildlife like the polar bear; it also has a direct impact on global warming. Melting ice drives another of the positive-feedback loops that are so common in the climate change world.

As we've seen, it's the Sun's energy that heats up the world. But not everywhere is equal when it comes to solar warming. The lighter in shade a surface is, the more energy is reflected back out to space (greenhouse gases permitting). The glittering whiteness of an ice sheet is ideal to flash back a good portion of the energy, while the dark waters of the ocean absorb significantly more. Water takes in more heat from sunlight than does ice. So the more the Arctic ice melts, the more energy is absorbed, melting even more ice—positive feedback. And even though the Arctic melting doesn't contribute directly to sea-level rise, because of this positive feedback, it does contribute further to global warming.

Much more worrying than the Arctic from the point of view of ocean levels is Greenland. If we think of Greenland at all, it tends to be either as a cold little place between Europe and America, or as the first example of dubious advertising, when it was optimistically given a name that implied verdant pastures in an attempt to attract gullible Norse settlers. But in climate terms, the interesting (and potentially frightening) thing about Greenland is its ice sheet.

More accurately, this is no mere ice sheet, but an ice mountain range. The Greenland ice sheet covers over one and a quarter million square kilometers (half a million square miles)—think Texas, California, and Florida combined—and is mostly over 1.6 kilometers (a mile) high. At its thickest, the ice sheet is nearly two kilometers (ten thousand feet) high, half the height of the tallest mountain in the United States, Mount McKinley.

According to NASA, through the 1990s the ice sheet was shrinking by around fifty cubic kilometers (twelve cubic miles) a year. That's a lot of ice—but it would still take between one thousand and ten thousand years for the Greenland ice sheet to melt completely. There's no room for sighs of relief, though. As Jim Hansen, director of the Goddard Institute for Space Studies and George Bush's top in-house climate modeler, graphically put it, "[Greenland's ice is] on a slippery slope to hell." By 2000, the rate the ice sheet was melting had accelerated so much that it was already losing vastly more than had been estimated just ten years before. The assumption had been that the ice would gradually melt from the surface downward, trickling its way to the sea as runoff water. But what is actually happening is startlingly different.

Lakes of water are forming on top of the ice sheets. These sheets aren't always uniformly solid. If there's a crack in the ice below a lake, the water can rush down, opening up the crevasse further as it flows until it has passed through the entire sheet to its interface with the rock below, where the water flow can eat away from beneath, enabling huge swaths of the ice sheet to float off the land. "[If] the water goes down the crack," says Richard Alley of Pennsylvania State University, "it doesn't take 10,000 years [to reach the base of the ice sheet], it takes 10 seconds." If the entire

Greenland ice sheet were to end up in the ocean, it would raise sea levels by seven meters (twenty-three feet). And this is without considering the impact of the melting of the Antarctic ice cap, which is on land, and so also contributes to the rise in sea level.

If the disappearing ice sheets weren't enough, there is also plenty of evidence that the glaciers around the world are also disappearing with unprecedented speed. Not only do these contribute to sea-level rise (the glaciers of Tajikistan alone hold eight hundred cubic kilometers, or two hundred cubic miles of water), but water from glaciers is essential for the irrigation and drinking water of many countries. Around 10 percent of northwestern China's water supply comes from glacier meltwater, for instance, and there are higher percentages elsewhere. Loss of glaciers will have a devastating effect on the economy and social well-being of a number of countries.

Sea-level rises are real and are happening. The Carteret Islands in the South Pacific are already being abandoned, their two thousand inhabitants displaced by the rising ocean. The current best guess suggests the islands will be totally submerged by 2020. Perhaps even more striking is the fate of Tuvalu, another collection of islands in the South Pacific, which forms a nation in its own right. The ten thousand people of Tuvalu are also having to give up their homeland. Before long that country will be a small modern-day equivalent of the mythical Atlantis, disappearing under the waters.

Many of the world's great cities are on coastlines and would have to be abandoned if sea-level rises reach a fraction of the levels that now seem entirely feasible. The timescale for this is uncertain. Conservative estimates put the rise by 2100 at half a meter (1.6

feet) but this is without the impact of positive feedback and the unexpected behavior of the Greenland ice sheet. The change in the Arctic perennial ice in 2007 was eighteen times faster than was predicted just ten years ago. By February 2007, sea-level rises were happening twice as fast as was predicted in 2001. Without a transformation in approach to climate change, the five meter (sixteen feet) mark could easily be reached in our lifetime. It would take only an extra 2.7 degrees Celsius (5 degrees Fahrenheit) to take the world to the conditions of the mid Pliocene, when sea levels were 80 feet higher than today. Imagine the New Orleans flood, but massively deeper and never abating. Cities like New York and London would not stand a chance.

Global warming will change the shape of the inhabited world. Over 20 percent of the world's population lives within thirty kilometers (twenty miles) of the coast, and the number of people living in these at-risk areas is growing at twice the average global rate. Rising seas would mean that most of the U.S. eastern seaboard would have to be abandoned, along with half of Florida, as would low-lying shore areas inhabited by hundreds of millions around the world. And this is not the limit. As we've seen, if the Greenland ice sheet melted, sea levels would rise seven meters (twenty-three feet.) The collapse of the fragile West Antarctic ice sheet would raise levels by up to another six meters (twenty feet), while the whole of the Antarctic ice cap melting would precipitate an extra sixty meter (two-hundred-foot) rise (though this is thought unlikely to happen with temperature rises of less than around 20 degrees Celsius (36 degrees Fahrenheit).

Any figures for sea-level rise also need to include the impact of storm surges. In some areas—around the New England coast, for example—when the storms are at their height there are expected to be sea-level rises of around three feet more than are otherwise predicted, well before the end of the century.

We might be dependent on energy coming in from the Sun, but in terms of matter, the Earth is largely a closed system. Extra droughts in some parts of the world mean more wetness elsewhere. As well as the impact of sea-level rise, some parts of the world can expect increased rainfall, and particularly more heavy storm rain. At the moment the increase is relatively slight—in 2001, the IPCC estimated increased precipitation of between 5 and 10 percent in the Northern Hemisphere over that of the previous hundred years—but there's more to come.

A significant fear is that global warming will produce more hurricanes like Katrina, the storm that devastated New Orleans and the surrounding coast in 2005. There is no certain evidence that climate change was behind the significant rise in numbers of hurricanes in 2005. As the oceans heat up, it should be easier for hurricanes to form, but there are other factors that come into play, and scientists are reluctant to commit themselves to saying that hurricane formation is on the increase. (This is a reassuring counter to those who think climate change scientists have a hidden agenda and make predictions that show that man-made climate change is responsible for everything that goes wrong with the weather.)

However, even if the increase in numbers of hurricanes in 2005 was a blip, it does seem true that there is a rise in the power of the type of tropical storm that can cause so much damage. Two studies in 2005 both showed that the energy levels of hurricanes is on the rise, with twice as many storms at the highest category 4 and 5 levels as were recorded in the early 1970s. There shouldn't be a similar effect with powerful tornadoes, though. The really big tornadoes are the product of a very special kind of thunderstorm that doesn't seem to be influenced by global warming. The smaller, more common tornados may be on the rise, but equally it could be that we are just noticing and reporting them more.

As things get worse, there will be huge disruptions to normal services. Availability of electricity, gasoline, and natural gas will be increasingly restricted as the need to respond to climate change goes critical. At the same time, with stocks of nonrenewable fuels running short and sources of supply becoming more remote, there is a growing opportunity for disruption of supply by natural disasters and terrorists. We could see a regular or even permanent breakdown of these services that are essential for our everyday lives.

Climate change also contributes directly to power outages. Extreme summer temperatures are often responsible for failures of power systems, in part because of the heavy load from air-conditioning, and also when power lines expand and sag, coming into contact with nearby trees and causing blackouts. The dramatic weather systems generated by global warming, including tsunamis and hurricanes, can also wreak havoc with power distribution systems. In December 2006 around 1.5 million homes in the states of Washington and Oregon were blacked out, some for up to a week, as power lines were brought down by howling windstorms and heavy rains. In January 2007, storms in the United Kingdom left 300,000 households without power, many for several days. As the impact of climate change grows, these will become very familiar headlines.

It doesn't help that the electricity grids of many countries are suffering from overload and age. As systems become more complex, their susceptibility to freak accidents and technical problems grows. In 2003 there were two large-scale electricity blackouts in Western countries. The Northeast blackout plunged a sizable part of Canada and the northeastern United States into darkness, leaving a total of 50 million people without power. The same year, an enormous power outage left the whole of Italy and parts of Switzerland without electricity, causing upheaval for a record 56 million people.

Of course, prediction isn't an exact science. We can't forecast the weather accurately for more than a few days out, so it seems optimistic to assume that we can know how the world's climate will change over tens of years. But while there will always be varying interpretations as long as there are different scientists analyzing the data, the consensus is now hugely in favor of global warming being a real, growing threat.

Some skeptics still point out that the changes so far have been relatively slight, and may not have a huge impact before the end of the century, but they are missing two significant points. First, the impact has begun. If you doubt this, speak to a citizen who has lost his home to unprecedented wildfires or coastal floods. Second, it's a mistake to assume that the change we see now will continue at such a relatively slow pace.

According to the Australian climate change expert Will Steffen, the world is not usually a place of gentle, slow drift. "Abrupt change seems to be the norm, not the exception," says Steffen. On twenty-three occasions during the last ice age, for instance, air temperatures went through massive climbs, pushing temperatures up by as much as 15 degrees Celsius (28 degrees Fahrenheit) in around forty years. Around half of the entire warming between the ice ages and interglacial periods that followed—again, changes on the order of a huge 15 degrees Celsius—took place in just ten years.

When the Earth undergoes major change it tends to be in sudden, large steps—this is something that is a relatively recent discovery. Richard Alley, in a report for the U.S. National Academy of Sciences, concluded, "Recent scientific evidence shows that major and widespread climate changes have occurred with startling speed. . . . This new thinking is little known and scarcely appreciated in the wider community of natural and social scientists and policymakers." We might be predicting a half meter (1.6-foot) rise in ocean level by 2100 (plus up to another 3 feet of storm surges) based on current, slow, steady rise, but we have to prepare for the possibility of a precipitate increase in temperature that will result in much faster rises in sea level.

Even without such a change, there is a real possibility that current predictions are underanticipating the impact because positive feedback may accelerate the process.

Sometimes, even our attempts to make the environment better can have an ironic and unexpected effect. Aerosols, the scientific term for suspensions of fine particles in the air, typical of much airborne pollution from smog to black smoke, are something that has been cut back significantly as we manage to clean up the air. But aerosols have a helpful effect where global warming is concerned. Unlike greenhouse gases, they stop the Sun's energy on the way in, and so have a cooling effect on the ground below. (This is reversed if soot particles, for instance, land from the aerosol on snow, darkening it and reducing reflection.) At the moment aerosols could be reducing the global-warming impact of greenhouse gases by up to half—but this contribution is liable to seep away as we achieve cleaner air.

Everyone from the most adamant denier of human-caused global warming to the greenest of scientists agrees that there's a lot of uncertainty in the predictions of just how much the effect will be. That's inevitable because they are dealing with a very complex and only partly understood system. Clouds, for example, have a big impact on the climate. Low clouds have a cooling effect; high clouds trap infrared radiation and warm us up. Different types of clouds make dramatically different contributions to heating or cooling. Attempting to include the feedback produced by clouds into climate models produces a huge range of variation.

This means that though the most likely predictions are still for a one or two degrees Celsius temperature rise within a century, it certainly isn't impossible to be looking forward to a rise of 8 to 12 degrees Celsius (15 to 20 degrees Fahrenheit)—plenty to make the most dire predictions of the impact of climate change a reality in our lifetimes. The knowledge that there's uncertainty doesn't mean we can just cross our fingers and hope the threat goes away. It's all the more reason to be ready, in case things head for the unpleasant end of the prediction range. And something we know for certain is that even if the predicted averages are true, we will experience worse, because as we have already seen, we don't experience averages; we live through the peaks and troughs. The damage is caused by the worst the weather can throw at us.

Is total devastation of society as we now know it inevitable from climate change? Thankfully, no. There are three possibilities that could save us.

The first is that it could all be a big mistake. Only twenty-five years ago some scientists were forecasting a return to ice age conditions rather than global warming. The Earth's weather system is immensely complex. We can't make detailed weather predictions more than a few days ahead. Beyond that, chaos reigns, as very small changes now can make a big difference to weather in the future. While the urban myth that a butterfly beating its wings on one continent can cause a hurricane on another is not true, chaos theory assures as that we will never be able to give a 100 percent accurate weather forecast over as short a period as a fortnight.

The models used to predict climate change are immensely complex, and subject to a lot of doubt. However, what we can do is take the best picture from a range of forecasts; this is how regular weather forecasts have been improved vastly over the last ten years. And doing that, the prognosis is not good. We can also compare the predictions of the models from a few years ago with what has happened since. So far, practically every model has been too optimistic. Things seem to be getting worse faster than the models predict.

The theory could still be wrong, and that could still save us. We may find that the world's climate surprises us and takes a whole different turn—but there is no good reason to assume that this is going to happen, and it's a pretty weak basis to plan the future of the world on.

The second possibility to mitigate the impact of climate change is that we make enough changes to the way we act, reducing greenhouse gas emissions and improving the way we soak up carbon dioxide, methane, and other such gases, to be able to hold off the kind of temperature rises that are currently predicted.

The global financial difficulties of 2008–9 could help a little with this, as there was a slowing down at least of the rate of rise of emissions, though some initiatives to try to overcome the recession, like those encouraging consumers to buy more new cars, would have had a negative effect on the environment.

Perhaps most encouraging are changes to both power generation and transport. On the power-generation side we have seen attempts to clean up power station exhaust gases, an increasing use of "renewable" resources like wind and wave power, and carbon cap-and-trade schemes to encourage organizations and governments to reduce emissions. For cars and trucks we have seen the May 2009 initiative, which promises leaner, greener automobiles in the future, reducing emissions by about one-third by 2016, with the equivalent impact on emissions of taking over 100 million cars off the road.

However, great though these actions are, most climate scientists tell us it's not enough to cut back on emissions. We need to get to a state where we're actively taking greenhouse gases out of the atmosphere. This is one of the possibilities for the third wave of fixes for climate change—where science is used to reduce greenhouse gas levels in an active fashion.

The most straightforward approach is to get hold of greenhouse gases and lock them away. This is what trees do—but unfortunately, they're much too slow at it for a tree-planting program to help with our short- to medium-term climate change problems. (And trees also have the problem that eventually they will die and rot, sending much of the carbon back into the atmosphere.)

We can already scrub the carbon out of the output of power stations, and with time could be able to do this with smaller emitters, like car engines. The carbon dioxide is captured, often by pushing it through solvents that latch onto it, then taken somewhere it can be stored long term. In theory this could just involve pumping CO_2 into the deep oceans; but it would gradually escape, and doing this would also contribute to increased acidification of seawater, putting corals and other marine life at risk.

More practical is pumping the gas into underground fissures and disused oil fields. Carbon dioxide is heavier than air, and with appropriate capping, such stores could keep the greenhouse gas locked away for thousands of years.

The trouble with much carbon capture, whether it's operating on car exhausts, your domestic boiler, or a power station, is that the processes involved can be energy intensive. The exhaust gases are bubbled through a liquid solvent that reacts with the CO_2, pulling it into the liquid. You then have to either wastefully and dangerously dispose of the solvent, or use a considerable amount of energy getting the carbon dioxide back out of solution so it can be stored away.

At the University of California, Los Angeles, a location all too familiar with car exhaust gases, scientists have been developing new carbon-capture materials. These zeolitic imidazolate frameworks are collections of tiny crystals that are like traps for CO_2. The crystals have pores in them that the carbon dioxide molecules can slip into, but find it difficult to get out of. Because the CO_2 has not undergone a chemical reaction, it can be extracted from the crystals by simply dropping the air pressure, leaving the crystals fresh to be reused. The team at UCLA hope that they will be able to test the crystals live in a power station within a year or two.

Others are looking at ways to recycle the carbon from the

atmosphere, actively reducing levels of greenhouse gases. There is a technique that can use solar energy to process carbon dioxide and hydrogen to produce hydrocarbons—the basic components of fuel oil and gas. Such carbon recycling could be used to actively reduce carbon levels, or simply to prevent them rising any higher by reusing the hydrocarbons produced this way, rather than burning fossil fuels.

These are the straightforward approaches that science can take, but others are more surprising, or more extreme. There are many suggestions; Here are three samples. One is to move from farming cattle and sheep to raising kangaroos. The reasoning here is that ruminants, grass-feeding mammals, are a major contributor to global warming. As an animal like a cow eats it burps up considerable amounts of methane. Although we don't hear as much about methane as we do about carbon dioxide, it is a powerful greenhouse gas, as we've seen, with around twenty-three times as strong an effect as carbon dioxide. The output of such livestock worldwide contributes 18 percent of all greenhouse gas emissions (in terms of impact)—more than all forms of transport combined. But kangaroos are different. They don't burp methane.

One possibility is to convert ranches to farming kangaroos, but a more likely approach is to look at what makes kangaroos different from cows and sheep. The kangaroos have a unique kind of bacterium breaking down plant matter in their stomachs. While the bacteria in cow and sheep stomachs pump out the greenhouse gas, the kangaroo equivalent doesn't. It has been known for some time that a diet rich in clover will partially suppress production of methane, but efforts are also being made to vaccinate against the offending bacterium, or even to try switching cows' stomachs over to the kangaroo bacteria.

A second idea that has been tried on a small scale is seeding the ocean with iron filings. These encourage the growth of algae, which take carbon dioxide from the atmosphere to build their cells, but (hopefully) don't release it back when they die, as they sink to the bottom of the ocean. On a large-enough scale an increase in algae levels would have a noticeable impact on greenhouse gas levels, but we just don't know what the result of dumping so much iron in the sea would have on other organisms, what the effect of a massive increase in algae populations would have on the marine ecology, or even how much the iron would really benefit the algae.

A final idea that is taken seriously in some quarters, despite seeming like a science-fiction fantasy, is to accept that there's nothing we can do about the level of greenhouse gases, and instead to reduce the level of natural heating of the Earth to compensate for the global warming coming from the greenhouse gases. The Earth's main source of heat is the Sun. So why not stop some of the sunlight from ever reaching the Earth?

The idea here would be to put an enormous screen in space that would cast a shadow over part of the Earth. This would have the immediate effect of reducing the Sun's heating power, it's true, but the cost and complexity of the idea are staggering. Ken Caldeira of the Carnegie Institution for Science of Washington, who has modeled the effects of a solar shield, has said, "We would need to be confident that we would not be creating bigger problems than we are solving. Therefore, it is important both to understand the mess we are in today—how close are we to making irreversible changes, how fast can we alter our energy system—and to understand what might happen should we try to avoid some of the worst outcomes by engineering our climate."

Scientists approach ideas to tweak the environment back the other way, away from global warming, with real trepidation. We have already demonstrated how good we are at messing up the environment, and we are all too aware of other situations where attempts to play nature at its own game have failed. Often, for instance, when a predator from a different country has been introduced to control a pest it has resulted in an out-of-control population of the predators, lacking the situations that normally keep them in check. The burgeoning population of predators then takes on prey they were never intended to attack, and throws the whole ecosystem out of balance.

Similarly, but on a much larger scale, there is a concern that taking action to reverse the effects of climate change could either produce unwanted side effects, or could go too far the other way, resulting in equally undesirable climate change involving global cooling. There's another, more subtle danger, too. If we give too much attention at this stage to possible future scientific solutions, we might be tempted not to do anything about cutting our emissions of greenhouse gases, assuming "we can just fix it"—when we have no evidence as yet that such a fix will ever be feasible.

What all now seem to agree on is that we can't afford to wait until things go horribly wrong before we start to try out different "geoengineering" approaches in small-scale trials. Without doing this, there is little chance of being able to assess the risk involved. Yet those trials that should be enabling us to get a better idea of what's feasible tend to come up against strong opposition from environmental groups, worried about the effects on local ecosystems where the trial is undertaken. We have to accept that we can't protect everywhere and everything. If we are to be ready to help reverse climate change, we need some short-term sacrifices to help us along the way.

Once more we see here the huge difficulty of dealing with a global issue, using actions that will have local effects. It's the inherent problem of that trite mantra "Think globally, act locally." When we act locally to influence a global issue, like climate change, the result will be not a global change but a variety of local changes. You can't practically turn down the temperature on the whole planet. Any geoengineering solution is likely to result in some areas dropping more in temperature than others—which will result in winners who get a better climate and losers who might experience drought.

The fact that impacts may vary by location has already come close to generating an international incident. In late 2005, Russian scientist Yuri Izrael attempted to persuade the Russian government to try releasing around 600,000 tons of sulfur particles into the atmosphere. Just as happens after a large volcanic eruption, this would result in reduced sunlight hitting the Earth and a reduction in global warming. However, it would be impossible to ensure that such an action would not cause droughts in some parts of the world.

This wouldn't necessarily be seen as just an experiment gone wrong. There are United Nations agreements in place prohibiting military (or other hostile) uses of environmental-modification techniques. This ban was brought in after U.S. attempts to use seeded rain to make terrain difficult to cross in the Vietnam War.

If such a particle-based sunshade did result in droughts or other weather conditions dangerous to life, particularly in a second country, it's quite possible it would be considered an act of aggression—though it might be hard to prove the source of the problem. It's clear that any attempt to engineer our way out of the problems of climate change other than by reducing emissions is fraught with difficulty.

All the initiatives to reduce our impact on the planet, and more like them, are great. And it's just possible that option one is true, and we don't have to do anything, because the models are wrong. But I am pessimistic. Even if the Western nations managed to become carbon neutral overnight, aspiring economies like China and India are continuing to increase their greenhouse gas outputs at an alarming rate. They are willing to talk about caps on emissions—but it is only natural that they feel they should be allowed to catch up a bit before they are set the same limits as those of us who already make huge contributions per head to global warming.

The other problem here is that almost all the solutions to climate change require long-term investments—but our whole political system is not set up to support long-term investment. Politicians like to see a quick fix, with plenty of bang for the buck. Climate change investment just can't be like that.

As we've seen, for instance, we need to invest a lot more in nuclear fusion. This technology would enable us to generate power with very low emissions—and without the reliance on rare minerals and the production of large quantities of radioactive waste that go along with traditional fission reactors. However there is only one new experimental fusion reactor planned *in the world*. The

United States, which you might expect to be leading such research, doesn't have a single experimental fusion reactor, and has reduced its level of funding to the ITER because the project is too long term.

The result of this short-termism is that the political action being taken to reduce the impact of climate change is too little, too late. It will help mitigate the impact—and every little bit helps—but I don't believe we will see real-world commitment to dealing with climate change until we have a state of near disaster in large parts of the globe.

It's depressing. I wish it were different. But it isn't. Green campaigners can jump up and down and predict doom as much as they like, but I think we should instead be thinking as much as possible about how can we mitigate the impact of climate change on human beings and our cities, because we aren't going to change our ways until things have gotten really bad.

One aspect of climate change that isn't always noticed is that the phenomenon is not a problem for the Earth. Green publicity material often accuses us of putting the planet at risk. But we aren't risking the Earth with our actions, not in any way. Our planet will cope just fine. We might end up making it temporarily uninhabitable for many of the living creatures we are familiar with, ourselves included, but to the planet itself there will be little change. What's more, bacteria will continue to thrive.

We tend to think of ourselves as the dominant species, but bacteria have been around a lot longer, and thrive in a much wider range of environments. In the next chapter we look at the uncomfortable relationship human beings and science have had with these smallest of living things, themselves possible bringers of mass destruction.

CHAPTER FIVE
EXTREME BIOHAZARD

||||||||||||||||||||||||||||||||||||||||||||||||||||||||||||

> *But however secure and well-regulated life may become, bacteria,*
> *Protozoa, viruses, infected fleas, lice, ticks, mosquitoes, and*
> *bedbugs will always lurk in the shadows ready to pounce when*
> *neglect, poverty, famine, or war lets down the defenses.*
> —Hans Zinsser (1878–1940), *Rats, Lice, and History* (1934)

At the end of 2002, a medical panic from Asia spread around the world. A new disease, severe acute respiratory syndrome (SARS), seemed about to devastate humanity. In a few months, there were thousands of cases in and around China, with outbreaks occurring in unexpected cities like San Francisco and Toronto as modern air travel provided the virus with an ideal vector for taking on the world.

As it happens, SARS proved relatively easy to contain, but a more justified concern emerged in Mexico in April 2009. Swine flu is a respiratory disease that originated in pigs and then spread to humans. It is caused by a specific strain of the influenza virus known as H1N1. These letters refer to the proteins that stick out from the flu virus's surface and are its mechanism for latching onto its victims. The H refers to hemagglutinin and the N to neuraminidase. Different variants of these two proteins mean that the virus will attach to different receptors in the victim's body. So, for instance, H1 viruses tend to bind in the upper respiratory tracts, making them easier to spread by coughing and sneezing, but often making them less severe than an H5 virus (like bird flu), which tends to bind in the lungs and bring on pneumonia.

The 2009 outbreak of swine flu is related to the common seasonal flu strain that kills thousands of people every year, but because of genetic variation it requires a different vaccination. On June 11, 2009, the World Health Organization declared swine flu to be pandemic, putting it at the top level of the WHO's alert structure, Phase 6. This means that it has caused sustained outbreaks in at least two countries in one region (the WHO divides the world into six regions, mostly but not all corresponding to continents), and in a third country in a different region. Once a disease is pandemic it is spreading around the world, and is impossible to contain.

At the time of writing more than seven hundred people had died from swine flu and more than 1 million Americans had been infected. By September 2009 cases were in decline, but they were expected to take off again as northern countries headed into winter conditions, where flu thrives. Swine flu is frightening because of that pandemic nature—its spread around the world is unstoppable. Thankfully, it appears to be a relatively mild variant compared to an earlier H1N1 pandemic. The so-called Spanish flu pandemic of 1918 is thought to have infected around 40 percent of the world's population and to have killed more than 50 million people.

A virus like flu is one of nature's weapons of mass destruction. In general, a biological weapon is one where the active ingredient is a dangerous infectious microorganism, or the toxins that such organisms produce. To give an idea of the potential for damage of biological weapons, it has been estimated that sixty pounds of anthrax, the material used in the postal attacks that followed 9/11, could kill a similar number of people to one of the nuclear weapons dropped at the end of the Second World War—between 30,000 and 100,000 people.

Although the creation of biological weapons is outlawed in most countries, this doesn't stop research on deadly bacteriological and viral agents, both for medical purposes and for defense against biological weapons. Research labs have the potential to release, accidentally or intentionally, a deadly agent that, like the flu, can spread around the world.

We will see that there's nothing new about the use of biological agents as weapons. But before exploring these most insidious of weapons of mass destruction, we need first to consider the less self-active agents that also address a biological weakness, which were deployed so horribly in the first half of the previous century: chemical weapons.

In a sense, all weapons interfere with the normal functions of the human metabolism—in the case of a bullet or a knife by crudely punching a hole in the body's delicate workings—but chemical and biological weapons disrupt the body's normal functions in a more indirect, and hence more scary, fashion. Arguably poison gas, with its indiscriminate killing and ability to sweep across a wide area, was the first true weapon of mass destruction to be deployed.

Although gas was probably used in poisoning well before the twentieth century, and had been proposed as a weapon as long ago as the time of Leonardo da Vinci, it had not been employed in a war, and was supposedly banned in the Hague Convention of 1899, which sought to place gentlemanly restrictions on the means for human beings to kill one another. But it soon became clear in the years following 1914 that the First World War was to be anything but gentlemanly, and both sides were prepared to try out gas if it meant a better chance of success in the dire battles of the trenches.

It can be argued (and has been) that there really is nothing different about the use of gas over any other weapon. Gas enthusiasts argue it's just another way to disable or kill the enemy, and can often be used to move an enemy out of the way without significant casualties. This was certainly put forward as a defense of its use at the time of the First World War. But it is not an argument that feels *right*. Many military personnel on both sides were privately disgusted by the horror of this silent, creeping killer. One German officer, reflecting on his army's early use of gas, commented, "Poisoning the enemy just as one poisons rats struck me as it must any straight-forward soldier . . . repulsive." It was the military equivalent of knifing someone in the back, rather than face-to-face.

It's true that gunfire and explosives can have horrendous effects on the human body—and don't always hit the intended targets—but at least such weapons provide a *directed* killing force. Gas, once released, acts under its own casual influences, be they weather or terrain, drifting, diverting, rolling forward indiscriminately. What's more, an individual bullet will kill one person. A shell might destroy tens, a conventional bomb hundreds. But gas can massacre thousands.

It seems likely that it was the repulsion raised by the idea of using gas that caused one extra and unexpected death during the

First World War. Clara Haber, wife of the chemist and gas warfare pioneer Fritz Haber, shot herself in May 1915. Although she left no note, all the evidence is that she was driven to take her life by her disgust with her husband's work. But her death would not stop him or his deadly program.

It's possible that Clara's suicide was in response to hearing about Haber's first battlefield success with gas. This took place near Ypres in Belgium, one of the place names ingrained into the minds of all those with family members who took part in the First World War, and known to the English-speaking troops as "Wipers." Algerian soldiers, part of the French army, were sheltered in rows of trenches opposing the German forces. The troops were locked in the slow, painful process of attack and retreat that typified the horrific muddy battlefields of that war.

On April 22, 1915, following an artillery barrage, the usual cloud of dust seemed strangely yellow-green in hue, and moved toward the troops with surprising vigor. But the Algerians saw nothing to fear, and carried on as normal. Minutes later, as the cloud continued to move forward and rolled into trench after trench, men were dying. The yellow-green cloud was chlorine gas, released from a chain of six thousand cylinders along the German front line and carried across to the French trenches by the prevailing wind.

Anyone who has handled swimming pool chemicals will be familiar with that burning in the nose and tickle in the throat that typifies the initial attack of chlorine—but this concentration of the gas produced symptoms on a wholly different scale. As the troops' eyes and mouth burned, a terrible coughing fit shook their bodies. The delicate lining of their lungs was being burned away, causing them to drown in the fluid that oozed out, leaving them frothing at the mouth. There is some dispute as to how many were killed by the gas, but it numbered in the thousands.

The power of the gas attack consisted in much more than its ability to kill. It spread fear. Winston Churchill, shortly after the First World War, emphasized this with enthusiasm. "I do not understand this squeamishness about the use of gas," he commented. He saw it as a positive that it "spread a lively terror." For most, though, this was not a positive. Men died in the trenches all the time, and on the whole, their comrades fought on alongside the bodies. A gas attack was different. Those who didn't die ran. The gas attack did not affect just the group of soldiers it disabled; it also cleared out the rest of the troops. One German soldier later commented after that first chemical offensive that they had been able to walk with their weapons tucked under their arms, just as if they were strolling along in a game hunt.

It was this ability to clear the battlefield of the enemy without necessarily killing them that made some insist that chemical weapons were, in fact, more humane than any other form of attack. As long as enough troops ran away or were partly protected by the limited gas masks of the time, the result was relatively few deaths, but many soldiers rendered incapable of fighting. This was put forward as a double benefit—not only was it less evil than killing outright; it also meant that the enemy had to expend more resources looking after those casualties. But few who experienced a gas attack could agree with the suggestion that this was a more humanitarian approach to fighting.

The Allies would also use gas within a few months, reducing any moral advantage they might have had from Germany's breaking of the Hague Convention. Chlorine, though devastating, was only the beginning of the chemists' excursion onto the battlefield. As more and more troops were provided with gas masks, scientists on both sides worked hard to ensure that they maximized the chances of exposing the enemy to the noxious gas. Extra materials were mixed into the cylinders to make the skin intolerably itchy or to induce sneezing, in an attempt to force the soldiers to remove their masks.

Although chlorine poisoned well enough, it gave the soldiers a fair amount of warning, both by being visible and from the initial burning sensations in the eyes and nose. Before long, Haber's German scientists had brought another chemical agent to the battle front—phosgene. A little more complex than the elementary chlorine, this is effectively a compound merging carbon monoxide and chlorine—$COCl_2$. Phosgene is invisible, and though it has a detectable odor, it smells pleasantly of new-mown hay, making it unremarkable in the countryside. There is no real warning of its deadly attack. Furthermore, phosgene has a cumulative effect, so an earlier dose can later be supplemented to reach a fatal level as the gas blocks the proteins that enable oxygen to be processed by the alveoli in the lungs, leaving the victim airless and dying.

Phosgene was soon followed into the trenches by the more complex and horrendous mustard gas, a compound of carbon, hydrogen, chlorine, and sulfur. This was deployed by the Germans in 1917 and, unlike its predecessors, did not disperse in hours, but could leave the battlefield uninhabitable for weeks, or even months. This is because mustard gas is really a liquid that can be sprayed like an industrial weed killer. It is extremely poisonous and causes terrible blistering on any exposed skin, both externally and internally, where it can wreak fatal damage.

The effects of mustard gas may not be felt for a number of hours after exposure, making it easy to build up a debilitating or deadly dose without being aware of it. The Germans made heavy use of mustard gas, deploying over a million shells filled with the substance in just ten days when it was first brought to the battlefield. Mustard gas is one of the few First World War chemical agents to have been deployed in recent times, when the Iraqis used it on Kurdish separatists in the 1980s.

The Allies were not slow to pick up on the new gas developments, and though lagging behind the Germans for most of the period when gas was used, by 1918 they were ready to attack back on a massive scale with chemical weapons. The assault was prevented only by the ending of the war. An American researcher, Winford Lee Lewis, had even produced a next-generation chemical weapon beyond anything the Germans had made, modestly called Lewisite, though thankfully this was not to be used in anger.

Lewisite was a compound of carbon, hydrogen, arsenic, and chlorine that beat mustard gas at its own game. Not only did it attack tissue, raising horrible blisters; Lewisite poisoned its victim on contact with the skin, so it wasn't even necessary to breathe it in. It was enough to be hit by droplets of this aerosol. Lewisite was capable of killing rapidly in very small concentrations.

The difficulty of controlling the spread of gases on the battlefield, and stronger conventions against their use, meant that chemical weapons would never again feature as significantly in warfare as they did in World War I. Yet, another poison gas developed by the same Fritz Haber would come to have even darker associations. This was the pest control fumigant Zyklon B. Releasing deadly hydrogen cyanide gas, the Zyklon canisters could kill in a confined space in seconds. Cyanide gas proved unsuitable as a battlefield weapon because it dispersed too quickly in the open air, but indoors it had a cold, killing efficiency. It would be deployed to horrendous effect in the gas chambers of the Nazi concentration camps.

The most modern class of chemical agents are nerve gases, which are usually split into kinds that act through breathing (and tend not to stay around for long), such as sarin, and those that act through contact with the skin (and are able to contaminate a site much longer), like VX. As is the case with many chemical agents, most nerve "gases" are actually liquids that are spread as an aerosol of fine droplets.

The majority of the nerve agents that attack through the respiratory system are called the G series, referring to their development in the late 1930s by German scientists. The skin-contact agents are the V series and were developed in the 1950s after a British scientist noticed the toxic effects on mammals of an organophosphate pesticide. These poisons are called nerve gases or nerve agents because they disrupt the nervous system, preventing control signals from being sent to the organs of the body and causing death as the body ceases to be able to activate the breathing mechanism. In effect, a nerve agent cuts the lines of communication within the body.

Considering just how well established is the concept of nerve agents as deadly weapons—most people have heard of them without being quite sure what they are—they have seen surprisingly little use in warfare. The Germans manufactured large quantities of the G series agents during the Second World War, going as far as producing artillery shells containing them, but never deployed them in battle. Although Britain developed the more potent V series agents in the 1950s, the British promptly, unilaterally gave up chemical weapons, including nerve gases, a move followed by many other countries. One of the G series nerve agents, GA or tabun, was used in a limited quantities by Iraq in the Iran-Iraq war, but the best-publicized use was the terrorist attack in 1995 using agent GB or sarin, on the Tokyo subway system.

During the morning rush hour of March 20, 1995, five members of the religious cult Aum Shinrikyo entered the Tokyo underground railway system, each carrying plastic bags containing a little less than a pint of liquid sarin. These bags were placed on the floor of the subway train and pierced using the bizarre mechanism of sharpened umbrella points, then left to spread the agent through the train and the tunnel as the motion of the train carried traces of sarin in the air through the carriage.

This was not a particularly efficient way of spreading the nerve agent—unlike a V series agent, sarin has to be breathed before it has an effect, and it doesn't evaporate particularly quickly at room temperature. Attacks on five separate subway lines killed twelve people and resulted in more than five thousand people going to the hospital, of whom around fifty were severely affected, and around one in five were more than trivially ill.

This Tokyo attack followed another terrorist incident carried out by members of Aum Shinrikyo the year before in the city of Matsumoto, where eight died. These incidents demonstrate a number of aspects of the use of chemical weapons. They show that a terrorist group with sufficient resources is perfectly capable of producing chemical weapons—but also that, despite the huge disruption and distress caused, nerve agents may have a relatively small impact when spread by improvised means.

Aum Shinrikyo's activities, incidentally, highlight that we should not rule out the ability of terrorists to access weapons of mass destruction because of the expense involved. Aum had a huge amount of cash in its coffers—some estimates have placed their assets at around $2 billion. Despite the apparently casual means of delivery, the nerve gas attacks on the subway were not undertaken on the cheap. The cult had set up a laboratory in Australia to develop the nerve agents and test them (mostly on sheep). According to some reports, the excess baggage costs alone of shipping laboratory equipment into Australia amounted to $300,000. The group also attempted (with less success) to buy old Soviet nuclear weapons.

The relative lack of impact to date doesn't make the deployment of chemical weapons by terrorists impossible, nor does it mean that chemical agents don't have the potential to be used in a deadly attack. Perhaps the biggest risk would be if some sort of aerosol system could be used to introduce a nerve agent into the air-conditioning system of a large office building.

There was awareness of the potential for terrorist use of chemical agents well before terrorism was recognized as a significant issue on the U.S. mainland. In 1989, then secretary of state George Shultz said, "Terrorists' access to chemical weapons is a growing threat to the international community. There are no insurmountable technical obstacles that would prevent terrorist groups from using chemical weapons."

Even so, all the evidence is that chemical agents will rarely be the weapon of choice for many terrorists or rogue states. However scary or dangerous chemical weapons are, they have less emotional impact than biological weapons. A chemical weapon may be indiscriminate once it is released, but at least it can be deployed solely on a battlefield. A biological weapon has the potential to spread without control—and in the long term will inevitably do more harm in the civilian population than the military. This makes biological agents particularly attractive to the terrorist.

What's more, we have a very sane, natural fear of being infected by serious biological agents. If there is such a thing as folk memory, plague will always lurk there as a dark reminder of our shared past. "Ring a ring o' roses" goes the old children's rhyme, speaking of the rash that accompanied plague. "Atishoo, atishoo! We all fall down." Dead. (There is some doubt about the link of the rhyme to the plague, but the sentiment works.)

It's well to remember when we contemplate even the horrendous casualties at Hiroshima and Nagasaki that in the four years from 1346 to 1350, the Black Death—which in all probability was bubonic plague—killed one-third of Europe's population. One in every three persons dead. Look around a busy office or at a bustling street, and imagine that happening now. Every third person dying. That would amount to around 90 million deaths in the United States alone. Biological hazards leave deep mental scars.

As a type of weapon, in an unsubtle form, biological materials have been in use for a long time. It has been a common practice since ancient times to use rotting corpses (animal and human) to make wells unusable for the enemy, and there are examples of biological weapons being applied more directly in medieval times.

Take the fate of Feodosia, a town in Crimea in the southern Ukraine. In the fourteenth century this was an Italian outpost, providing a trading center for dealing between the Italian city of

Genoa and the exotic East. In 1346, the town's walls were surrounded by a Tatar army, determined to remove what were seen as
invaders from their territory. Their timing was poor—the campaign coincided with the start of the great outbreak of the Black
Death. Soon, the besieging army was weakened as soldiers went
down with the ravaging of the plague. But rather than accept this
as a sign that they should retreat, the Tatars turned their illness to
their advantage.

They began to catapult dead bodies over the town walls into
Feodosia. Of itself this is a horrible, demoralizing act that would
have caused fear among the inhabitants. But their intention was
much darker. These were the bodies of plague victims. Before long,
the Black Death had broken out inside the fortified town. The surviving Italians evacuated the trading post, heading back to their
own country and leaving Feodosia to the Tatars. They had been
the victim of a crude biological siege weapon.

Between the fourteenth century and the U.S. War of Indepen-
dence, everything from diseased corpses to smallpox-riddled blan-
kets were used in this way, but biological weapons have rarely been
used in modern times, partly because of the difficulty of controlling their impact and partly because of a fear of retaliation. Perhaps the most dramatic, if crude, attempt to use them was during
the Second World War, when Japanese planes dropped porcelain
containers over northeastern China. These fragile bombs contained millions of fleas carrying the plague virus.

Another, disputed example from the Second World War demonstrates the difficulties faced by those attempting to use a bacteriological weapon on the battlefield. In 1942, a German tank attack
on the Russian front was abandoned when the soldiers faced an

outbreak of the disease tularemia, sometimes called rabbit fever. Although the outbreak was initially primarily among German troops, it then spread to thousands of Russians, both military and civilian. Although there is no definitive evidence that this outbreak was engineered, as is believed by the Russian biological weapons expert Ken Alibek, it demonstrates the way that a biological attack on the battlefield can easily overtake both sides of the conflict.

But the difficulty of controlling the spread of diseases from the deadly agents has not stopped work from being done on the production and delivery of biological weapons, taking them far beyond the crude infectious capability of a disease-ridden corpse or a flea bite. Nor does this lack of direction make biological weapons any less appealing to terrorist groups, who see the horror raised by the uncontrolled nature of a biological agent as a positive rather than a negative attribute.

As we have seen, the Russians may well have been making use of biological weapons by 1942, and we know that the United Kingdom and Canada began work on them in 1940, while the United States started its own program in 1943, largely because of suspicions that the Germans seemed to be adding a collection of deadly biological agents to their armory.

The unit set up in the United States to work on the weapons, given the low-profile name War Research Service to conceal its activity, had its central base at Camp Detrick in Frederick, Maryland, where the usual suspects of deadly infectious diseases from plague to typhus were worked on, with a particular focus on anthrax. Thousands of pounds of biological bombs were produced at the Detrick site but were never used during the Second World War.

After the war, some had hoped that the biological-warfare

capability of the United States, referred to as a "dirty business" by Secretary of War Henry Stimson, could be wound down, but the discovery of advanced biological weapons in Japan led instead to the development of an escalating series of weapons that were just as much an attempt to keep ahead of the enemy as anything that was happening on the nuclear front. It wasn't until 1969, under President Richard Nixon, that the biological warfare program was abandoned, due to a combination of public pressure and military doubts about the viability of biological weapons on the battlefield.

In looking at what biological weapons can do, we need to be aware of the different kinds of agents. Most of us know that, for example, a bacterial infection (which will usually respond to an antibiotic) is different from a viral infection (which won't). Similarly, there are bacterial and viral types of biological weapon, plus fungal infections (most familiar in the everyday world in the harmless athlete's foot) and rickettsias. These last are less familiar disease producers, somewhere between a bacterium and a virus, often found on fleas and ticks, responsible for illnesses like typhus (different from typhoid, which is bacterial).

The biological agent we probably hear most about—certainly one of the most feared—is anthrax. "Anthrax" originally meant a carbuncle, or malignant boil, reflecting one of the possible symptoms of an attack of the bacterium. It is primarily a disease of livestock, and its actions have been recorded since antiquity. Anthrax can take effect by ingestion or through the skin, but the main concern about its use as a weapon is anthrax's ability to infect through inhalation. Spores carrying the bacterium *Bacillus anthracis* are breathed in by the victim. Once the disease has taken hold there is a 90 to 95 percent chance of death.

To begin with, the victim of an anthrax attack feels as he would with a cold—a blocked nose, a cough, some aching of the joints. Once the symptoms become apparent, usually a day or so after infection, it is too late to treat the victim. Despite a brief period when the symptoms subside, the bacteria is rampaging through the body via the lymph nodes, spreading a toxin that will impact all the organs, but particularly the lungs, which fill with liquid. By then, the skin will have taken on a blue tinge and breathing will become intensely painful, leading to a fatal choking spasm.

One of the reasons anthrax is a preferred biological weapon with the military is that it doesn't spread from person to person (although spores from corpses can cause infection); rather, its spores kill on being breathed in—this makes it more controlled than an infectious disease, and hence better for a military action.

Another benefit of anthrax to the military is that it is easy to produce, can be stored for years as a dry powder without losing its potency, is cheap, and is easy to disperse in aerosol form (or, as the 2001 anthrax attack demonstrated, as a powder in an envelope). Anthrax will also stay potent for a long time, making it an effective way to render an area contaminated and unusable. When the British army tested anthrax as a weapon on Gruinard Island off Scotland in 1942 (it was never deployed in battle), it took nearly fifty years before the island was safely decontaminated.

There are many other possibilities for a biological weapon, ranging from diseases like Ebola fever to the incredibly deadly natural toxin ricin, which is derived from the castor bean plant. Ricin is particularly scary because the lethal dose is so small— about half the size of a grain of sand—but it lacks some of the practical capabilities of the preferred biological weapons.

Other diseases that were frequently added to twentieth-century biological stockpiles were plague and smallpox, while the Russians also experimented with Legionnaires' disease—the illness sometimes caught from infected air-conditioning systems—and AIDS. Neither of these proved practical to make in a stable form that could be used in weapons, and AIDS has such a long incubation period that it wasn't suitable for a military attack, though its potential for striking fear was huge, hence the attempt. Some have suggested, however, that terrorists might prefer an apparently low-risk biohazard: foot-and-mouth disease.

Foot-and-mouth is primarily a disease of animals with cloven hooves. It can be caught by human beings, but the result is only a mild fever and some blistering—it is not a serious problem for humans. However, as the outbreak in the United Kingdom in 2001 demonstrated, foot-and-mouth is highly contagious, and can result in huge disruption to livestock management and the general ability to move freely around a country. Strangely, the approach typically taken to control foot-and-mouth seem to have been caused by an upper-class desire for perfect-looking animals, and is not a logical response to the infection.

In the days when the aristocracy was still in control of Britain, it was fashionable for a landowner to keep an attractive herd of cattle on his estate—they were considered a visual asset, like having a beautiful lawn or an ornamental building in the garden. Foot-and-mouth rarely kills cattle, nor does it make their meat inedible, but it does disfigure them—they don't look as pretty afterward. So draconian measures were instituted to try to prevent a relatively harmless disease from spreading. It's arguable that those measures, still in place today, are more damaging and disruptive than the disease itself, though admittedly it can reduce milk output and cause sterility. But there is now an effective inoculation against foot-and-mouth, so the disruption that the disease causes is arguably no longer necessary.

Worrying though the problems foot-and-mouth can generate are, it has more potential as a weapon of mass *disruption* than a weapon of mass destruction. This is, certainly, a terrorist aim in its own right. Disruption is a powerful propaganda tool, and disruption caused by many terrorist threats (such as the security clampdown after the discovery of a plot to take liquid explosives onto aircraft) does have a direct impact on our ability to live freely. Yet disruption in itself does not have the Armageddon quality that is the focus of this book.

Compared to many weapons of mass destruction, biological weapons are relatively easy to make. Often, given a starting colony, the biological agent will make itself—and the technology involved can be little more sophisticated than the equipment found in an industrial kitchen, provided there are appropriate mechanisms in place to stop the agent from escaping and attacking the workers.

In at least one recorded case, a terrorist succeeded in obtaining deadly bacteria simply by placing an order with a medical supplier. As it happens, Larry Harris, a member of a white supremacist group called Aryan Nations, was too impatient, and caused suspicion when he kept calling the supplier to ask why his plague bacteria had not arrived—but it is chilling that he was able to simply order a sample of plague over the phone and have it delivered by FedEx. It's to be hoped that since September 11, procedures checking those placing orders for deadly diseases have become significantly more robust.

The ease of production or access is likely to continue to make biological agents attractive to terrorists. And there is evidence of more than just a hypothetical threat. The year 1972 saw the arrest in Chicago of members of a terrorist group called the Order of the Rising Sun. They had in their possession seventy-five pounds of typhoid bacteria cultures, with which they planned to contaminate the water supply of the cities around the Great Lakes. In 1984, members of another fringe group succeeded in spreading typhoid bacteria via the salad bar in an Oregon restaurant—many diners became ill, though none died.

Similarly, rogue states can produce biological weapons relatively easily. There are many more facilities already existing worldwide to produce biological agents than there are for the more high-tech weapons of mass destruction. Any laboratory developing or producing harmless and much needed vaccines against disease can easily be turned into a factory for manufacturing biological weapons.

There are, however, two key problems that the manufacturer of these weapons has to confront: how to get the disease agent into a suitable form to use it as a weapon and how to deliver it. In their natural form, many bacterial agents are easily destroyed by heat, by ultraviolet light, or simply by being kept on the shelf too long. As Russia's Ken Alibek has commented, "The most virulent culture in a test tube is useless as an offensive weapon until it has been put through a process that gives it stability and predictability. The manufacturing technique is, in a sense, the real weapon, and is harder to develop than individual agents."

This process involves not only "growing" the agent itself, but rendering it in a form that makes it easy to store—often a dry, powderlike material produced by blasting the agent with powerful jets of air—and mixing it with additives that will help preserve it. It might also be necessary to cover the agent in tiny polymer capsules to protect it from light where it is liable to be damaged by the ultraviolet component of sunlight. The biological weapons business has as much in common with the packaged-food industry as it does with the weapons trade.

Then there's the problem of delivery. The selected bacterium, virus, fungus, or rickettsia may be deadly enough when the disease is caught, but how do you deliver the agent to your target? We might mostly think "natural is best" when choosing food these days, but the natural means for these bacterial agents to spread can be slow and are almost always difficult to control.

When the Soviet Union was at the height of its biological warfare program it tested the effectiveness of different means of spreading biological agents on an unknowing population. Harmless bacteria like *Bacillus thuringiensis,* known as simulants, would be spread over a populated area using different delivery mechanisms to see how the technology performed. The test subjects would never know that they were being exposed to a disease, and the monitoring would be under the cover of routine medical examinations.

With an agent like anthrax that can be manufactured in powder form it is relatively easy to deliver the biological weapon to its target by spreading the powder from the air, or from an explosive container, but many biological agents are liquids. These will usually be dispersed into the air as tiny droplets, turning the liquid into an aerosol.

Although a liquid bacteriological agent can be dispersed from the air using low-energy bombs (if the bomb is too powerful it tends to kill the agent), there is a much easier solution, which is as readily available to the terrorist as to the rogue state. Most of the technology that is used to spread pesticides and other agrochemicals can be used equally well to launch a chemical or biological attack. This is particularly true with crop-spraying aircraft, which are ideal for dispersing a deadly agent over an occupied area.

Western countries had given up their biological weapons by the 1970s, and it was thought that this was also true of the Soviet program, until revelations came from defectors in the 1990s—it now seems that the USSR kept a heavy-duty biological weapons manufacturing process in place all the way up to the early 1990s.

According to Ken Alibek, a former senior manager in the Soviet program who has now defected to the West, in the 1980s, Biopreparat, the organization responsible for much of the Soviet biological weapons development, was coming up with a new biological weapon every year—either enhancing an existing threat like anthrax to make it more resistant to antibiotics, or weaponizing a whole new disease.

At the time, the Soviets genuinely believed that the United States was lying about having given up biological weapons, and they considered it essential to make their biological warfare program more and more extreme to maintain an imagined lead in a race that didn't exist. As Alibek says, "We were engaged in secret combat against enemies who, we were told, would stop at nothing. The Americans had hidden behind a similar veil of secrecy when they launched the Manhattan Project to develop the first atomic bomb. Biopreparat, we believed, was our Manhattan Project."

This is a hugely revealing comment. Not only was the Manhattan Project a vast and secret enterprise; it was intended to produce the weapon that would end the war—and the weapon that would prove the ultimate lever in dealing with the enemies of the United States at the time. It is not fanciful to suggest that the Soviet hierarchy thought that having a significant lead in biological weapons would give them a similar potential lever over their American rival.

At the peak of the program, the Soviets had the same long-range, multiple-warhead SS18 missiles that carried nuclear weapons prepared to carry enough anthrax to be capable of taking out a whole city like New York. At the same time, they were developing technology to release much smaller canisters of biological agents from cruise missiles, which would have had the advantages of stealth and much more accurate targeting, essential for effective use of biological weapons.

While many countries have worked on biological weapons, most now shy away from this despised means of attack, whether on moral grounds or because their militaries are not happy with the indiscriminate nature of the technology. However, just as rogue states like Iraq have used chemical weapons in the relatively recent past, so such states are still likely to consider using biological agents.

In principle, both biological and chemical weapons are covered by conventions, but these political controls have proved less effective than their nuclear counterparts because they don't have the same accompanying regime of international inspection. It's also much harder to spot a biological or chemical test than a nuclear test, and there are many more ifs and buts in the biological conventions to allow for research on the prevention of disease. If you are producing enriched uranium there is really only one thing you can do with it, whereas research into viruses and bacteria is at the heart of the production of new medical cures.

On the chemical side, as we have already seen, was the Hague Convention of 1899, which was strengthened in 1907 to prevent the use of all "poison or poisoned arms." But this had very little influence on either side in the First World War. The most recent attempt to prevent the use of chemical weapons is the Chemical Weapons Convention of 1997, which has been ratified by all but a handful of countries (significant omissions include North Korea and Syria, with Israel signed up to the convention but not yet ratifying it at the time of writing). This treaty, like the nuclear one, does have an inspectorate in the form of the Organisation for the Prohibition of Chemical Weapons (OPCW), based in The Hague, Holland. But there is a significant get-out clause that allows for research related to protection against chemical weapons, which many would argue is a good defense for developing such weapons.

Things are even less well covered on the biological side. The Biological and Toxin Weapons Convention was set up earlier, in 1975, and has rather fewer states committed to it, many of the absent countries being in Africa. However, this convention has no monitoring body—it is merely a statement of intent—and as the history of development of biological weapons in the USSR shows, it is a convention that some states have been prepared to flout. Sadly, a significant part of the resistance to having a verification process comes from the United States, where the biotechnology industry has lobbied hard not to have any biological equivalent to the OPCW's role for chemical agents.

Meanwhile, the United States maintains laboratories working on biological agents to research ways to detect, resist, and counter biological attacks. This proved justified in 2001, when it was proved that terrorists were capable of both considering and delivering biological weapons. The anthrax package attack killed five people, infected seventeen more, and had a large impact in terms of inconvenience and cost as systems were put in place to prevent the attacks from being repeated.

However, biological agents seem not to be the weapons of choice of terrorist groups like al Qaeda. This lack of interest is probably caused by a combination of a preference for the immediacy of explosives with a cultural dislike of the concept of biological warfare. Even so, we should not be surprised if biological weapons are used again. Most likely targets would be those where movements of air naturally provide for the spread of the agent.

One way to achieve this is in air-conditioning systems. These already have a tendency to spread the natural biological agent Legionnaires' disease, and they would be equally effective at spreading other diseases if the agents were injected into the system appropriately. The same technique could work on airliners, though the potential target population is much smaller. Or, insidiously, a powder-based agent like anthrax could be seeded in subways, where the wind produced by the trains would spread the agent through the system.

Our agencies need to remain vigilant to the dangers of biological attack. The outbreak of an engineered plague, designed by human intervention to be difficult to resist, is a nightmare possibility. Yet it is less exotic than a newer threat to the human race that has emerged from scientific discoveries: nanotechnology.

CHAPTER SIX
GRAY GOO

||||||||||||||||||||||||||||||

> *These microscopic organisms form an entire world composed of*
> *species, families and varieties whose history, which has barely*
> *begun to be written, is already fertile in prospects and findings*
> *of the highest importance.*
>
> —Louis Pasteur (1822–95), "Influence de M. Pasteur sur les
> progrès de la chirurgie," Quoted by Charles-Emile Sedilliot,
> paper read to the Académie de Médecine (March 1878)

Pasteur's words in the quote that opens this chapter refer to the "microscopic organisms" of nature. But imagine the construction of man-made creatures on an even smaller scale, an army of self-replicating robots, each invisible to the naked eye. Like bacteria, these "nanobots," endlessly reproducing devices, could multiply unchecked, forming a gray slime that swamped the world and destroyed its resources.

Each tiny robot would eat up natural resources in competition with living things, and could reproduce at a furious rate. This sounds like science fiction. It is—it's the premise of Michael Crichton's thriller *Prey*. But the idea of working with constructs on this tiny scale, nanotechnology, is very real. It has a huge potential for applications everywhere from medicine to engineering, from sunblock to pottery glaze—but could also be one of the most dangerous technologies science could engage in, as the so-called gray goo scenario shows (gray goo because the nanobots are too small to be seen individually, and would collectively appear as a viscous gray liquid, flowing like a living thing).

Louis Pasteur and his contemporaries didn't discover microorganisms. The Dutch scientist Antoni von Leeuwenhoek peered through a crude microscope (little more than a powerful magnifying glass on a stand) in 1674 and saw what he described as "animalcules"—tiny rods and blobs that were clearly alive, yet so small that they were invisible to the naked eye. This idea of a world of the invisible, detectable only with the aid of technology, was boosted into a central theme of physics as atomic theory came to the fore and it was accepted that there could be structures far smaller than those we observe in the everyday world.

The original concept of the atom dates all the way back to the ancient Greeks, though, if truth be told, it proved something of a failure back then. The dominant theory at the time was taught by the philosopher Empedocles, who believed that everything was made up of four "elements": earth, air, fire, and water. It was the kind of science that seemed to work from a commonsense viewpoint. If you took a piece of wood, for instance, and burned it, the result was earthlike ashes, hot air, fire, and quite possibly some water, condensing from the air. And these four "elements" do match up well with the four best-known states of matter: earth for solid, water for liquid, air for gas, and fire for plasma, the state of matter present in stars and the hottest parts of flames.

This theory would be the accepted wisdom for around two thousand years. By comparison, the alternative idea, posed by the philosophers Democritus and his master, Leucippus, was generally considered more a philosophical nicety than any reflection of reality. Democritus proposed cutting up a piece of matter repeatedly until it was smaller and smaller. Eventually you would have to come to a piece that, however fine your knife, was impossible to cut further. This would be indivisible, or in the Greek *a-tomos.* An atom.

These atoms of Democritus were not quite what we understand by the term today. Each different object had its own type of atom—so a cheese atom would be different from a wood atom— and the shape of the atom was determined the properties of the material. Fire, for instance, would have a sharp, spiky atom, where water's was smooth and curvaceous. Yet in this largely forgotten concept there was the seed of the idea that would blossom in the early nineteenth century, when British scientist John Dalton devised the modern concept of atoms as incredibly tiny particles of elements, building blocks that would be combined to make up the substances we see around us, either in the pure elementary form or interacting with different atoms to make compound molecules.

Dalton was led to this idea by work a couple of decades earlier by the French scientist Antoine-Laurent Lavoisier, who has the rare (if hardly desirable) distinction among scientists of being executed, though this was for his role as a tax collector at a time of revolution, rather than for his theories. Lavoisier laid down the basics of modern chemistry, showing how the same quantities of different substances always combined to make specific compounds. It seemed to imply some interior structure that made for these special combinations.

Yet the existence of such tiny objects as atoms was only grudgingly accepted. As late as the early twentieth century there was still doubt as to whether atoms really existed. In the early days of atomic theory, atoms were considered by most to be useful concepts that made it possible to predict the behavior of materials without there being any true, individual particles. It was only when Einstein began to think about a strange activity of pollen grains that it became possible to demonstrate the reality of the atomic form.

In 1827, the Scottish botanist Robert Brown had noticed that pollen grains in water danced around under the microscope as if they were alive. To begin with, he put this down to a sort of life force that was driving the tiny living particles in motion. But he soon found that ancient and decidedly dead samples of pollen still showed the same activity. What's more, when he ground up pieces of metal and glass to produce small enough particles—things that had never been alive—exactly the same dance occurred.

This was considered an interesting but insignificant effect until 1905, the year when Albert Einstein would publish three papers that shook the scientific world. One was on special relativity; the second was on the photoelectric effect, the paper that helped kick-start quantum theory; and the third was on Brownian motion. Einstein proposed that this random dance of tiny particles like pollen was caused by millions of collisions with the molecules— simple collections of atoms—that made up the water in which the grains floated.

The reality of the existence of atoms and molecules was confirmed with certainty only in 1912 by French physicist Jean Perrin, who took conclusive measurements that backed up Einstein's theory. And in 1980, most remarkably of all, Hans Dehmelt of the

University of Washington succeeded in bringing an individual atom to the human gaze. More accurately, this was an ion—an ion is an atom with electrons missing, or extra electrons added, giving it an electrical charge—of barium.

Just as the antimatter traps described in chapter 2 work, the ion was held in place by electromagnetic fields. The ion's positive charge responded to the field rather in the same way that a magnet can be made to float over other magnets, though the ion had to be boxed in by several fields to prevent it from flying away. Incredibly, when illuminated by the right color of laser light, the single barium ion was visible to the naked eye as a pinprick of brilliance floating in space.

Once we have the idea that everything from a single water molecule to a human being is an assembly of atoms, differing only in the specific elements present and the way those atoms are put together, a startling possibility emerges. If there were some way to manipulate individual atoms, to place them together piece by piece as a child assembles a Lego construction, then in principle we should be able to make anything from a pile of atoms. Imagine taking a pen or a hamburger and analyzing it, establishing the nature and location of each individual atom present. Then with suitable technology—we'll come back to that—it should be possible to build up, from ingredient stores of each atom present, an exact duplicate of that item.

But the proof of atoms' existence that came in 1912 didn't mean that it was possible to do anything with them in practice. Admittedly, in one sense, ever since human beings started to manipulate the world around us, we have been reassembling atoms. Whether simply chipping off bits of stone to make an ax head, or smelting metal and molding a tool, we were recombining atoms and molecules in new forms. But this approach was much too crude to enable any form of construction step by step with the fundamental building blocks.

Even now, more than a century after Einstein's paper, the most common way to manipulate atoms and molecules directly is crudely using accelerators and atom smashers. Science fiction's best guess of how we could handle such ridiculously small items was that we would have to be shrunk in size. So in Asimov's movie and novel *Fantastic Voyage,* for example, we saw miniaturized humans interacting with the microscopic components of the human body. The idea of being able to manipulate objects on the nanoscale seemed unreal.

Until recently, this prefix "nano" was familiar only to scientists. It was introduced at the eleventh Conférence Générale des Poids et Mesures in 1960, when the SI (Système International) units were established. As well as fixing on which units should become standards of measurement—the meter, the kilogram, and the second, for example—the conference fixed prefixes for bigger and smaller units from tera (multiply by 1 trillion) to pico (divide by 1 trillion). The penultimate prefix was nano (divide by 1 billion), derived from *nanos,* the Greek word for a dwarf. It's one-billionth, so a nanometer is a billionth of a meter (about 40 billionths of an inch), a truly tiny scale.

It was the great American physicist Richard Feynman who first suggested, in a lecture he gave to the American Physical Society in 1959, that it might become possible to directly manipulate objects at the molecular level. Feynman was a trifle optimistic. He said, "In the year 2000, when they look back at this age, they will wonder why it was not until the year 1960 that anyone began seriously to move in this direction." In practice we are only just getting there in the twenty-first century.

There are three huge problems facing anyone attempting to manipulate atoms to produce a new object. First is mapping the structure of an object—having an accurate blueprint to build to. Second is the sheer volume of atoms that have to be worked on. Imagine we wanted to put together something around the size and weight of a human being. That would contain very roughly 7×10^{27} atoms: 7 with 27 zeroes after it. If you could assemble 1 million atoms a second, it would still take 3×10^{14} years to complete. That's 300 trillion years. Not the kind of time anyone is going to wait for a burger at a drive-through.

And finally there is the problem of being able to directly manipulate individual atoms, to click them into place, like so many Lego bricks.

Feynman envisaged overcoming the problem of scale by using massively parallel working. It's like those old problems they used to set in school tests. If it takes one man ten hours to dig a hole, how long would it take a gang of five men? Feynman envisaged making tiny manipulators, artificial "hands," perhaps first just one-fourth of normal size. He imagined making ten of these. Then each of the little hands would be set to work making ten more hands one-sixteenth of the original size. So now we would have one hundred of the smaller hands. Each of those would make ten of one-sixty-fourth scale—and so on. As the devices got smaller, the number would multiply, until we would have billions upon billions of submicroscopic manipulators ready to take on our challenge.

Twenty-seven years after Feynman's lecture, author and entre-preneur K. Eric Drexler combined the "nano" prefix with "tech-nology" in his book *Engines of Creation* to describe his ideas on how it would be possible to work on this scale. He referred to the Feynman-style manipulator as an assembler, a nanomachine that would assemble objects atom by atom, molecule by molecule.

A single assembler working at this scale would take thousands of years to achieve anything. As we have seen, there are just too many molecules in a "normal"-scale object. To make anything practical using assembly would require trillions of nanomachines. Drexler speculated that the only practical way to produce such an army of submicroscopic workers would be to devise nanoma-chines that could replicate like a biological creature, leading to the vision of gray goo and the potential devastation portrayed in Crich-ton's *Prey.*

Before examining the realities of the gray-goo scenario, there are other aspects of nanotechnology that need to be considered. Working on this scale doesn't necessarily involve anything so complex as an assembler. One very limited form of nanotechnol-ogy is already widely used—that's nanoparticles. These are just ordinary substances reduced to particles on this kind of scale, but because of their size they behave very differently from normal ma-terials. The most common use of nanotechnology currently is in sunscreens, where nanoparticles of zinc oxide or titanium dioxide are employed to protect us from the sun's rays, allowing visible light to pass through, but blocking harmful ultraviolet rays. In fact, we have been using nanoparticles for centuries in some of the pigments in pottery glazing, without realizing it.

There is also a form of atomic manipulation and construction at the heart of every electronic device, and especially a computer like the one I have written this book on. We casually refer to "silicon chips," a weak, dismissive term for integrated circuits that totally understates what a marvel of technology these are. Using atomic layer deposition, one of the most advanced of the techniques used to "print" the detail on top of the base silicon wafer of a computer chip, layers as thin as one-hundredth of a nanometer can be employed. This is true nanotechnology.

Soon to be practical, with vast potential, are more sophisticated nano-objects—nanotubes and nanofibers. Often made of carbon, these molecular filaments are grown like a crystal rather than constructed and have the capability to provide both superstrong materials (as an extension of the current cruder carbon fibers) and incredibly thin conductors for future generations of electronics. Semiconducting nanotubes have already been built into (otherwise) impossibly small transistors, while carbon nanotubes could make one of the more remarkable speculations of science fiction a reality.

Writer Arthur C. Clarke conceived of a space elevator, a sixty-two-thousand-mile cable stretching into space that could haul satellites and spacecraft out beyond the Earth's gravity without the need for expensive and dangerous rocketry. Bradley Edwards, working for the NASA Institute for Advanced Concepts, commented in 2002: "[With nanotubes] I'm convinced that the space elevator is practical and doable. In 12 years, we could be launching tons of payload every three days, at just a little over a couple hundred dollars a pound." Edwards was overoptimistic—there is no sign of the development of a space elevator as we near the end of his twelve-year period—but nanotubes do have huge potential and will be used more and more.

All these simple nanostructures do have the potential to harm human beings. Because of the physics of the very small, the tiny particles or structures can react in unexpected ways—for instance, penetrating the skin and slipping through biological membranes that normally keep foreign matter out, or by being carried into the lungs to cause damage. We have seen in the past how much damage asbestos fibers—once treated as if they were harmless—can do. It seems only reasonable to take care when dealing with nanoparticles and other nanostructures.

Some of the early reactions to nanoparticles have, however, demonstrated massive ignorance. In January 2008, the Soil Association, the biggest organic certification body in the United Kingdom, banned nanoparticles from organic products. But in doing so, the Soil Association specifically banned only *man-made* nanoparticles, claiming that natural ones (like soot) are fine because "life has evolved with these."

This totally misunderstands the threat we face from a nanoparticle. A nanoparticle is most likely to be dangerous because of its scale, because the physics (rather than chemistry) of particles of this size is quite different from that of the objects we are familiar with. Where this is the case, that danger is just as present whether the particle is natural or it isn't. Even where scale isn't the only risk factor, natural nanoparticles can be dangerous because of the way they act or their ability to interact with the body. Viruses are natural nanoparticles, and like soot, they aren't ideal for the health.

The Soil Association defends its position by saying that its approach parallels a sensible attitude to carbon dioxide in the air, where there is no problem with the natural carbon dioxide, only the man-made contribution. This is a specious argument, both because carbon dioxide is carbon dioxide, and if levels are too high it doesn't matter where the molecules are coming from, and because there is no comparison between CO_2 and a nanoparticle that could be directly physically dangerous to humans.

To make matters worse, the Soil Association also says that it can't control natural nanoparticles present in the environment. They're just there. However, this is not relevant—the Soil Association isn't an environmental control group. Its role is to control what goes into organic products, and there is nothing to stop a manufacturer from putting natural nanoparticles into a product either by accident or intentionally. You might as well say we don't mind a manufacturer putting salmonella into organic food, because it's natural. If the Soil Association believes nanoparticles are a bad thing, it should ban all nanoparticles from a product that gets its seal of approval, not just artificial ones.

It is when they try to summarize their argument that the Soil Association lets slip the reason it takes this strange attitude. "The organic movement nearly always takes a principles-based regulatory approach, rather than a case-by-case approach based on scientific information." In other words, theirs is a knee-jerk reaction to concepts, rather than one based on genuine concerns about the dangers of nanoparticles. It is all about words like "natural" and "artificial," not about the nanoparticles themselves. In practice, we need to be concerned about all types of nanoparticles.

There are genuine and sensible worries about potential health risks from nanoparticles. As we have seen, because of their unique physics, they can penetrate barriers that stop other contaminants, whether these are natural barriers like skin or simple breathing masks. We do know it makes sense to minimize our exposure to breathing nanoparticles, and we should make sure there is long-term testing of the effects of any nanoparticles in substances we apply to our skins or ingest. But it seems unlikely that nanoparticles could cause worldwide devastation.

Moving up a step of complexity, we get to nanoscale machines, the sort of mechanical "hands" on the scale of molecules that Feynman described. Much of the work that has happened so far in this field has come not out of engineering shops as Feynman envisaged, but biology labs. There's a good reason for this: because the biological world, from the complex chemicals that operate within our bodies to stand-alone nanoscale entities like viruses, is replete with molecular machines operating at just the level we are considering.

Take proteins, the workhorse molecules that carry out the instructions of DNA in living cells. Proteins don't just carry signals by plugging into other molecules, or act as reinforcements in cell-based structures like cartilage. What are often long, string-like molecules fold into specific shapes—and the way a protein folds will determine how it then acts. It is arguable that because of the mechanical action of folding, such proteins are the simplest form of biological machines. We are used to thinking of machines as complex devices like a car or an iPod—but we should remember that basic devices like the lever, the pulley, and the screw are all machines in the technical sense.

Whether or not you class a simple folding protein as a machine, there are certainly machine functions in living creatures that are powered at the level of nanotechnology, whether it's the kind of ratchetlike "grab and pull" that proteins undertake in a muscle, multiplied millions of times to make your legs or arms move, or a truly complex machine like the flagellum found on some bacteria. These microscopic propellers, beloved of the supporters of intelligent design, have an ion-powered motor and a rotary socket. However, it's hard to see how these particular machines can cause real damage. They don't represent the kind of threat we are looking for in a nanoscale Armageddon.

For real devastation, we probably need to be looking at nanoscale robots—nanobots—which remain the ultimate aim of many nanotechnologists. Thinking back to Richard Feynman's vision, to be able to manufacture items atom by atom, we had three issues: mapping an object to know what to build, being able to work on enough atoms at a time, and being able to manipulate individual atoms. This last problem would require as a starting point special nanoscale machines—assemblers. Exactly how these would work is not clear, but they would effectively be nanobots, invisibly small robots whose sole role in life would be to take atoms and reassemble them in a new form.

To make this possible, we would need to manufacture these nanobots and to power them, and we would have to be able to give them the instructions they need to make whatever we require. While Feynman's "small hands, making smaller hands, making smaller hands" could be seen as a way to kick-start the process, inevitably the only practical way to make nanobots would be to have them capable of making themselves. That's because we would need not just a few million nanobot assemblers, but countless billions.

Remember the numbers for assembling an object around the size and weight of a human being. We would need 300 trillion such assemblers to achieve the task in a year at a rate of a million actions per assembler per second. We could achieve this only if, in effect, assemblers could breed.

Assuming for a moment that we had these self-replicating nanobots, they wouldn't have to turn into the sort of voracious monsters portrayed in Michael Crichton's book in order to become a threat to humanity. One possibility is that humans would become so dependent on nanobots that, should the technology fail, the human race would be doomed.

It's just possible this would happen if all manufacturing were replaced by assembler production lines. Imagine products being assembled molecule by molecule so anything and everything could be constructed by a single device. If such technology were perfected and cheap, it's hard to see why any conventional form of manufacturing would remain in use. Then, once we had become totally dependent on them, the failure of nanobots would cripple our technology-dependent society.

The impact on our lives could be much greater than just losing manufactured goods, many of which are luxuries we don't need to survive. Imagine a food-production device like those on the TV show *Star Trek: The Next Generation,* where food is assembled to order. If all our food were assembled instead of grown, then the world would face starvation if nanobots stopped working. But there is another possible way to become totally dependent on the technology. We could get to a stage where our bodies needed nanotechnology to survive from moment to moment.

Those like Ray Kurzweil who imagine a future where human

beings can effectively live forever believe that we will reach a point where we inject nanobots into our bloodstreams to fix our cells, and to replace entirely many of the functions of fallible human organs like taking over pumping blood from the heart. In principle an assembler can make anything, unstitching atom from atom and reconstructing the building blocks of nature into anything from a scarf to a TV to a piece of beef. There's no reason why its role shouldn't include making (or acting as) more efficient parts of a human body. If we became totally dependent on these nanobots to stay alive, then any large-scale failure of the technology could spell disaster for the human race.

One way this could happen is if nanotechnology suffered a failure that paralleled the problems affecting the natural nanomachines in our bodies. If we are to achieve Kurzweil's dream of tiny intelligent machines inside our bodies that keep us alive, we probably have to go some way down the assembler route. The only sensible way to build such nanobots is to use other nanomachines. We would then be susceptible to failures in the nanobots that resemble the biological problems that plague real living, reproducing creatures.

It's the replication process that produces a threat. As we already know, things can go wrong with replication in the natural world—when mutation occurs. If something is being replicated many, many times, there is a chance of an error in the copying. Usually that error will result in failure, but occasionally it can make a change that will make the replicating creature better.

Natural selection will ensure that the "better" form of the creature thrives, assuming it can pass on its difference, and eventually it will take over from the earlier version that lacks the enhancement. That's evolution in a nutshell. It happens with the biological machines that populate the world, and it could happen to nanomachines. Once machines have the ability to replicate, and to pass on changes in design, they can evolve, or pass on a fatal flaw to the whole population.

We'll come back to what might go wrong due to mutation when we look at gray goo, but before that we need to consider a different kind of failure. Perhaps the biggest danger facing a world dependent on nanotechnology would come not from the accidental evolution of a nanobot, but from the intentional handiwork of a hacker: a nanotechnology virus.

Every day in my e-mail inbox, I receive, besides the genuine e-mails I want, a whole host of others I don't want. There is spam, trying to sell me Viagra or encourage me to provide my bank details so I can receive a huge bequest. But worse still, there are e-mails carrying viruses, trojans, and worms, all hoping to take over my PC or cause damage. Some virus writers produce them for fun or as an intellectual challenge, but others are, in effect, electronic terrorists who hope to cause disruption and confusion. The result of this relentless impact from would-be attackers is that I have to have three programs running all the time—antivirus, a firewall, and antispyware, constantly battling to protect my computer.

These electronic vandals don't stick to a single technology. As long as it's widespread enough to be worth their attention (the reason Apple computers are relatively unscathed), they will get involved. Now that many cell phones are powerful pocket computers, virus writers have spread their attentions to this technology. In June 2004, the first cell phone virus emerged into the wild. One of the particularly unnerving things about cell phone viruses is that they behave in a way that's more like the real thing than anything that arrives on your computer.

The use of the term "virus" when referring to a malicious computer program has always had the potential to cause confusion. When the public first became aware of computer viruses, I was running the PC department of a large company, and once had a phone call from a worried executive who had recently become pregnant. She was worried about catching the computer virus and the danger of it causing damage to her unborn child. Cell phone viruses don't put humans at direct risk any more than computer viruses, but they certainly can hit our wallets, and they spread in a worryingly natural manner.

Here's a typical scenario featuring an attack by the phone worm Commwarrior. You are in a bar and need to make an urgent call. Your cell phone beeps—it asks if you want to accept a Bluetooth connection from someone you don't know. Sensibly, you click No, as you don't want to connect to a stranger. But before you can do anything else, up comes the request again. And again.

It keeps coming so fast that you can't place your call. So you finally say Yes just to get it out of the way—and with that Yes, your phone is infected. The Commwarrior virus has jumped from a stranger's phone to yours. Because of the way Bluetooth works, cell phone viruses and worms that use it jump from phone to phone when they are in close proximity. Your cell phone literally catches a bug like this by being near a phone that is infected. And once your phone is infected, the virus has the potential to start siphoning cash from your account.

It would be naïve to think that hackers who can jump on the

bandwagon so effectively with smart phones wouldn't try to do the same with nanomachines. If anything is going to give us cause to pause and think whether or not we want to go down this route, it is the possibility of hacking. Yet it's reasonably easy for this to be avoided. Computer viruses, whether on a PC or on a phone, are just programs. It's entirely possible to make intelligent electronic devices that can't have a program run on them other than the one that is built in.

We normally allow reprogramming of electronic equipment because software often needs updating and we don't want to throw the hardware away if the program is wrong. But nanobots exist in a different kind of world. They don't need to be reprogrammed—we can literally rebuild them molecule by molecule instead. Any programming is hardwired in the atoms. This is an antihacking advantage, and it makes the software less complicated.

With nonprogrammable nanobots we can avoid hacking. What isn't possible, though, is to prevent malicious people from corrupting the "seed" devices from which the nanobots will be produced; nor can we stop rogue technologists from producing their own nanomachines that can cause damage. But the genie is already out of the bottle. Just as it was impossible to forget the possibility of atomic weapons once the concept had been devised, we can't go back to a time when the idea of nanobots hadn't occurred to anyone. We are already at the stage of "the good guys had better build them, or it will only be the bad guys doing it."

And so we come back to the gray-goo scenario. In self-reproduction we have got a plausible mechanism to achieve the volume of nanobots required (provided we overlook the lack of a basic means to construct them in the first place), but we still need to have the "blueprints" of what the assemblers are going to build, we need to convey instructions to these assembler machines, and we need to provide them with the raw materials and power they require.

Where the object to be constructed is simple and repetitive, the instructions to the assembler could be input by hand. For example, a diamond consists of a set of carbon atoms in a simple, well-understood repeating pattern. All that is required to establish a true diamond with its familiar characteristics of hardness, transparency, and luster is the exact measurements of that repeating pattern. It would not prove difficult to set this up, just as we might provide instructions to a lathe to construct a chair leg.

Things become harder, though, if our assemblers are going to build a complex structure like a computer, or an organic substance, like a piece of roast beef. Here, Eric Drexler has suggested, we would need to have disassemblers, the mirror counterpart of assemblers. The role of these devices would be to swarm over an object and strip away its atoms, layer by layer, capturing the information needed to reproduce such an item. In the process the original object would, inevitably, be vaporized.

The obvious route for feeding the instructions from disassemblers to the assemblers that would make the new version of the vaporized object would be to learn from nature. In the natural world, chemical and electrical signals are used to pass control messages from biological machine to biological machine. Similarly, a sea of nanobot assemblers would have to receive instructions from their surroundings, whether by chemical, electrical, acoustic, or light-based means. One potentially valuable source of guidance here is in the control mechanisms used by superorganisms.

For centuries we have been fascinated by the way that apparently unintelligent insects like bees, ants, and termites manage collectively to undertake remarkable feats of engineering and mass activity. What has gradually come to be realized is that, for example, the bees in a hive are not really a set of individuals somehow managing to work together. Instead, the whole colony is a single organism.

When seen from that viewpoint, their capabilities and strange-seeming actions make much more sense. It is no longer difficult to understand why a particular bee might sacrifice itself for the colony. In the same way, in the human body, cells often "commit suicide" for the good of the body as a whole, a process known as apoptosis. Similarly it's less puzzling how the bees manage the kind of organization, hive building, and harvesting activities they do. If you consider each bee as more like a single cell in a multicelled organism, it is much easier to understand how the colony functions.

Although assemblers would have to be simpler than bees or ants, as they would be built on a much smaller scale, they could still make use of the mechanisms that a colony employs to communicate and to control the actions of individual insects. Bees, for example, pass on information in "waggle dances," and use chemical messages to orchestrate the behavior of the many individuals that make up the superorganism. In designing a swarm of assemblers, we would have similar requirements for communication, and might well employ similar superorganism techniques.

When it came to raw materials and power, we would have to enable the assemblers to take atoms to use in construction from somewhere nearby—they could hardly run down to the nearest store and fetch their materials—and would need to have them consume energy from sunlight, using photochemical or photoelectric means, or from chemical energy by burning fuels, just as most biological creatures do.

Given those possibilities, a number of gray-goo scenarios emerge. Imagine rogue assemblers, where the instructions have become corrupted, just as the instructions in our DNA routinely become corrupted to form mutations (and each of us has some of these flaws in his or her DNA). Instead of being told to use, say, a pile of sand—and only that pile—as raw materials, the assemblers could choose whatever is nearby, whether it's a car, a building, or a human being.

Similarly, if we opt for chemical fuel rather than light-based power, it's not hard to imagine a swarm of runaway assemblers that are prepared to take any carbon-based material—crops, animals, human beings—and use them as the fuel that they digest. And bearing in mind that assemblers have to be able to produce other assemblers—in effect, to breed—it's not difficult to imagine a sea of the constructs, growing ever larger as it consumes everything before it.

Then there is the terrifying possibility of an out-of-control swarm of disassemblers. The good news here is that disassemblers couldn't breed. To enable them to both construct and disassemble seems unnecessarily complex—disassemblers would be the product of a sea of assemblers. So a swarm of disassemblers would not be ever growing and self-reproducing as a group of assemblers could be. But disassemblers could, without doubt, cause chaos and horrendous damage, acting like a cartoon shoal of piranhas, stripping all the flesh from a skeleton. Only disassemblers wouldn't be limited to flesh—they could take absolutely *anything* and everything apart, completely destroying a city and every single thing in it.

However, these dangers are so clear, even at this very theoretical stage, that it's hard to imagine the situations where rogue assemblers could get out of hand being allowed to come close. It would be easy enough, for example, to avoid assemblers eating food (or humans) by limiting them to solar power; the leap to chemical energy sourcing would be too great for any realistic mutation (and unlike with biological mechanisms, we can build in many more checks and balances against mutation occurring). Similarly, disassemblers would be easy enough to limit, by giving them either limited lifespans or an aversion to various substances, reducing the threat they pose.

Furthermore, it's also worth stressing again that we may never be able to build effective nanomachines. There have been experiments producing promising components—for example, nanogears assembled out of molecules, and nanoshears, special molecules like a pair of scissors that can be used to modify other molecules—but just think of how much further we have to go. Not only do we need to build a complex mechanism on this scale, but we will have to give it a power source, a computer, and the mechanism to reproduce. Of these, the only one we can manage at the moment is to build the computer, and that is on normal macro scales, rather than something so small we can't see. (Yes, we have power sources, but battery technology isn't transferable to the scale of a nanomachine. There isn't room to carry much fuel on board, so the nanobot would need to harvest energy.)

Although there have been predictions of self-building (in effect, reproducing) robots for a long time, the reality lags far behind the hype. Back in March 1981, a NASA scientist announced on CBS Radio news that we would have self-replicating robots within twenty years. We are still waiting. And this is just to have something that can reproduce itself on the normal macro scale, without all the extra challenges of acting on the scale of viruses and large molecules.

Even if the technology were here today, there would probably be scaling issues. Just as you can't blow a spider up to human size because of scaling, you can't shrink something that works at the human scale down to the nano level and expect it to function the same way. Proportions change. The reason we will never have a spider the size of a horse is that weight goes up with volume, so doubling each dimension produces eight times the weight. But the strength of the legs goes up with cross-section, so doubling each dimension only produces four times as much strength. Weight gets bigger more quickly than the strength of a creature's limbs. The monster's legs would collapse.

Similarly, different physical influences come into play when working on the scale of nanotechnology. The electromagnetic effects of the positive and negative charges on parts of atoms begin to have a noticeable effect at this level. A quantum process called the Casimir effect means that conductors that are very close to one another on the nanoscale become powerfully attracted to one another. At this very small scale, things stick together where their macro-sized equivalents wouldn't. This could easily mess up nanomachines, unless, like their biological equivalents, they resort to greater use of fluids. Almost everyone making excited predictions about the impressive capabilities of nanomachines wildly underestimates the complexity of operating at this scale.

But let's assume we do develop the technology to successfully make nanobots. How likely is the gray-goo scenario? This envisages a nanomachine that has, thanks to a random error, got out of control, producing a "better" machine from its own point of view, though not for humanity, as it simply duplicates itself and consumes, becoming a ravening horde that destroys everything from crops to human flesh. It's a ghoulish thought. Yet the parallel with biology isn't exact. The big difference between a machine and a plant or an animal is that the machine is designed. And the right design can add in many layers of safeguard.

First there is the resistance to error. Biological "devices" are much more prone to error than electronic ones. Yes, there could still be a copying error in producing new nanobots, but it would happen much less often. Second comes error checking. Our biological mechanisms do have some error checking, but they aren't capable of stopping mutation. It is entirely possible to build error checks into electronic devices that prevent a copy from being activated if there was an error in the copying. Depending on the level of risk, we can implement as many error checks as we like. There is a crucial difference between design—where we can anticipate a requirement and build something in—and the blind process of evolution.

Third, we can restrict the number of times a device can duplicate itself, as a fallback against rampant gray goo. Finally, we can equip devices with as many other fail-safes as we like. For instance, it would be possible for nanomachines to have built-in deactivators controlled by a radio signal. All these layers of precautionary design would make any nanobots much less of a threat than those who find them frightening would suggest.

However, it doesn't really matter how clever our precautions are if those who are building the nanotechnology are setting out not to produce something safe and constructive, but rather something intentionally destructive. It's one thing to feel reasonably comfortable that we will not be threatened by the accidental destructive force of nanotechnology, but what happens if someone actively designs nanotechnology for destruction?

According to Friends of the Earth, "Control of nanotech applications for military purposes will confer huge advantages on those countries who possess them. The implications of this for world security will be considerable. The military uses of nano scale technology could create awesome weapons of mass destruction." It's important we understand just how much truth there is in this, and why scientists are so fascinated by nanotechnology.

In looking at military uses of nanotechnology, we need to build up to Friends of the Earth's supposed weapons of mass destruction, because there can be no doubt that nanotechnology itself will make its way into military use. Probably the earliest military application will be in enhanced information systems—the ability to drop a dust of sensors, practically invisible to the onlooker but capable of relaying information back to military intelligence.

Equally close down the line is the use of nanotechnology in special suits to support soldiers in the field, potentially using a whole range of technologies. There could be ultrastrong, lightweight materials constructed from nanotubes, exoskeletons powered by artificial muscles using nanotechnology, and medical facilities to intervene when a soldier is injured, using nanotechnology both to gain information and to interact with the body.

But to envisage the weapons of mass destruction that Friends of the Earth predicts, we need to return to variants of the gray-goo scenario, driven not by accident but by intent. Each of the most dramatic of the nanobot types of attack—disassemblers, assemblers using flesh as raw materials, and assemblers using flesh as fuel—has the potential to be used as a dramatic and terrifying weapon.

There is something particularly sickening about the thought of an unstoppable fluid—and swarming nanobots are sufficiently small that they will act like a liquid that moves under its own volition—that flows toward you and over you, consuming your skin and then your flesh, layer by layer, leaving nothing more than dust. It's flaying alive taken to the ultimate, exquisite level of horror.

This is a weapon of mass destruction that is comparable to chemical or biological agents in its ability to flow over a battlefield and destroy the enemy, but with no conventional protection possible—whatever protective gear the opposing forces wore, it could be consumed effortlessly by the nanobot army. It is a truly terrifying concept. But I believe that is an unlikely scenario.

The first hurdle is simply achieving the technology. I would be surprised if this becomes possible in the next fifty years. As was demonstrated by the NASA scientist who envisaged self-replicating robots by 2001, it is very easy to overestimate how quickly technology can be achieved in such a complex and challenging area.

Second, there is a matter of control. It is unlikely that a nanobot attack could be deployed in the same way as a chemical or biological weapon, as it could consume the friendly army as easily as the enemy. It would need to be deployed more like a nuclear weapon, provided the nanobots were programmed to have a limited lifespan—otherwise it's not clear why the destruction would ever stop.

Finally, provided the enemy has similar technology, some form of countermeasures become possible. If nanobots are programmed not to attack friendly personnel by using chemical scents or special radio signals, these protective measures could be duplicated to misdirect them. It's also possible to envisage a suit with outer layers of protective nanobots which feed on the attacking nanobots, neutralizing the attack.

The result of such defensive technology is that nanobot weapons would tend to fail to destroy the military, only massacring innocent citizens. This is a form of weapon—worse than indiscriminate—that is likely to be strongly legislated against if it ever becomes a reality, just as biological and chemical weapons are today. This doesn't mean that they would never be used. Rogue nations and terrorists could deploy them, just as they have other illegal weapons. But the international community would act to suppress their use.

It is fascinating how much interest nanotechnology has raised in the scientific community, out of all proportion to the benefits that the concept of working on this scale has delivered at the time of writing. The key technologies of the twenty-first century are often identified as GNR—genetics, nanotechnology, and robotics. Of these, only genetics has yet to have any real significance, unless you are prepared to count integrated circuits as nanotechnology.

Nanotechnology has such an appeal to scientists in part because it is often the path taken by nature. We may build on the macro scale, but many natural "machines" depend on workings at a subcellular level. For example, the power sources of the cells in our body, mitochondria, are just a few hundred nanometers across, yet they are vital to our existence. And then there is the fantasy (at least as we currently should see it) of Eric Drexler's assemblers in *Engines of Creation*.

If we could get assemblers to work, we would have, in principle, the opportunity to make almost anything at hardly any cost over and above the price of the raw materials. We would have a technology that could extend human life almost indefinitely, and that could enable us to haul satellites into space for a fraction of the current cost, or to analyze materials to a whole new level of subtlety. This is a stunning possibility that will keep nanotechnology in the minds of those who direct the future of science funding, even while the actual benefits currently being delivered by nanotechnology are much more mundane. But we have to bear in mind just how far out in the future this is. It is entirely possible that there won't be significant nanotechnology of the kind described in *Engines of Creation* this century.

Nanoscale robots are very much an idea for the future, but we don't have to look any distance ahead to see how misuse of another technology—information technology—could pose a real threat to our present-day world.

CHAPTER SEVEN
INFORMATION MELTDOWN

||||||||||||||||||||||||||||||||||||||||||||||||||||||||||||||||||||||

> *It has been said that computer machines can only carry out the*
> *processes that they are instructed to do. This is certainly true in*
> *the sense that if they do something other than what they were*
> *instructed then they have just made a mistake.*
>
> Alan Mathison Turing (1912–54), *A. M. Turing's*
> *Ace Report of 1946 and Other Papers,*
> ed. B. E. Carpenter and R. W. Doran (1986)

In 2005, a rat and a clumsy workman managed to bring down all of New Zealand's telecommunications, causing nationwide chaos. Science has given us a superb capability to exchange information, far beyond anything that nature can provide, but our increasing dependence on electronic networks leaves us vulnerable to large-scale destruction of data. In 2008 we saw how relatively small upsets in the banking system can cause devastation in the world economy—but this upheaval is trivial compared to the impact of the total loss or the wide-scale corruption of our networks of global data. The financial world would be plunged into chaos—and so would business, transport, government, and the military.

Such a science-based disaster might seem relatively harmless—it's only a loss of information—but our complex, unnatural society can exist for only a short time without its information flows. Without the systems that enable food to be provided to stores, without the computer-controlled grids that manage our energy and communications, we would be helpless, and the result would be an end to familiar, comfortable life, and could result in millions of deaths.

Let's go back to New Zealand in June 2005. Two small, totally independent acts came together to disrupt a nation. On North Island, a hungry rat chewed its way through a fiber-optic cable, one of the main arteries of electronic communication in the country. At about the same time, a workman was digging a hole for a power line in a different part of the country and sliced through a second cable. Although the communication system has redundancy, giving it the ability to route around a network failure, two major breakdowns were enough to close the stock exchange and to knock out mobile phones, the Internet, banking, airlines, retail systems, and much more. For five hours, the country was paralyzed by these twin attacks.

There is always the potential for accidental damage, such as happened in this New Zealand failure—but there is a much greater risk from intentional harm. The chance of such a dual hit on communications lines happening by accident is very small, but terrorists would consider it the obvious approach to achieve maximum effect. Governments are now taking the terrorists of the electronic frontier much more seriously than they once did. We have to accept that computer scientists, from the legitimate technologist to the undercover hacker, have the potential to present a real danger to society.

The reason that intentional damage presents a higher level of risk is that an accident will typically strike in one spot. It was only due to bad design and an unlikely coincidence that the rat and the workman managed to bring down the New Zealand network. But malicious attackers can make use of the interconnectedness of computers to spread an attack so that it covers a wide area of the country or even the world, rather than being concentrated in any one spot. The very worldwide nature of the Internet that provides so many benefits for business and academia also makes it possible for an enemy to attack unpredictably from any and all directions.

Although the threat seems new, it is a strategy that goes back a surprisingly long way—well before the time when most of us had computers—and we need to travel back in time to the earliest days of the Internet's predecessor to see how this all started.

We're used to the Internet as a complex, evolving hybrid of personal, commercial, and educational contributors, but it all started off with the military, and specifically ARPA. ARPA (the Defense Department's Advanced Research Projects Agency, later called DARPA) was set up in 1958 as a direct response to the panic resulting from the Soviet Union's launch of Sputnik, the first artificial satellite, the previous year. ARPA was established to ensure that the United States would lead the world in future high-technology projects.

From the early days, ARPA had funded computers in a number of universities and wanted a way to enable a user with a terminal in, say, Washington to be able to log on to a computer in California without having to travel to the specific site to use it. It was from this desire to log on to remote computers that the ARPANET was first established. But it soon became clear that the network, and the "packet-switching" approach that was adopted as its communications protocol, could also be used for computers to connect to one another, whether to pass data among programs or to allow the new concept of electronic mail to pass human messages from place to place.

Eventually a part of the ARPANET would be separated off to form MILNET, where purely military unclassified computers were sectioned off on their own, and the remainder of the ARPANET would be renamed the Internet, forming the tiny seed of what we now know and love. The Internet wouldn't really take off until 1995, when it was opened up to commercial contributors—initially it was the sole domain of universities.

In 1988 there were around sixty thousand academic computers connected via the ARPANET. The vast bulk of these were medium to large computers, running the education world's standard operating system, UNIX, though some would have been using the proprietary operating systems of companies like the then highly popular Digital Equipment Corporation (DEC) minicomputer manufacturer. In late 1988, operators running some of the computers noted that their machines were slowing down. It was as if many people were using them, even though, in fact, loads were light. Before long, some of the computers were so slow that they had become unusable. And like a disease, the problem seemed to be spreading from computer to computer.

To begin with, the operators tried taking individual computers off the network, cleaning them up, and restarting them—but soon after reconnecting, the clean machines started to slow down again. In the end, the whole ARPANET had to be shut down to flush out the system. Imagine the equivalent happening with the current Internet: shutting down the whole thing. The impact on commerce, education, and administration worldwide would be colossal. Thankfully, the ARPANET of the time was relatively small and limited to academia. But its withdrawal still had a serious cost attached.

This disastrous collapse of a network was the unexpected and unwanted side effect of youthful curiosity, and the typical "let's give it a try and see what happens" approach of the true computer enthusiast. The ARPANET's combination of a network and many computers running the same operating system seemed an interesting opportunity to a graduate student at Cornell University by the name of Robert Morris.

Morris's father, also called Robert Morris, worked for the National Security Agency (NSA) on computer security. But student Morris was more interested in the nature of this rapidly growing network—an interest that would bring him firmly into his father's field, with painful consequences. Because the ARPANET was growing organically it was hard to judge just how big it was. Morris had the idea of writing a program that would pass itself from computer to computer, enabling a count of hosts to be made. In essence, he wanted to undertake a census of the ARPANET—a perfectly respectable aim. But the way he went about it would prove disastrous.

Although what Morris created was referred to at the time as a computer virus, it was technically a worm, which is a program that spreads across a network from computer to computer. Morris had noted a number of issues with the way UNIX computers worked. The ubiquitous sendmail program, which is used by UNIX to transfer electronic mail from computer to computer, allowed relatively open access to the computers it was run on. At the same time, in the free and easy world of university computing of the period, many of those who ran the computers and had high-level access had left their passwords blank. It proved easy for Morris to install a new program on someone else's computer and run it. His worm was supposed to spread from computer to computer, feeding back a count to Morris.

There's no doubt that Morris knew he was doing something wrong. When he had written his self-replicating worm, instead of setting it free on the ARPANET from Cornell he logged on to a computer at MIT and set the worm going from there. But all the evidence is that Morris never intended to cause a problem. It was because of a significant error in his coding that he wound up with a criminal record.

When the worm gained access to a computer, its first action would be to check whether the worm program was already running there. If it was, there was no job to do and the new copy of the worm shut itself down. But Morris realized that canny computer operators who spotted his worm in action would quickly set up fake versions of the program, so that when his worm asked if it was already running, it would get a yes and would not bother to install itself. Its rampant spread would be stopped, and it wouldn't be able to conduct a full survey of the network.

To help overcome this obstacle, Morris added a random trigger to the code. In around one in seven cases, if the worm got a yes when it asked if it was already running, it would install itself on the computer anyway, and would set a new copy of itself in action. Morris thought that this one-in-seven restriction would keep the spread of his worm under control. He was wrong.

The Internet, like the ARPANET before it, is a particular kind of network, one that often occurs in nature. Because it has a fair number of "hubs" that connect to very many other computers, it usually takes only a few steps to get from one location on the network to another. What's more, it was designed from the beginning with redundancy. There was always more than one route from A to B, and if the easiest route became inaccessible, the system software would reroute the message and still get it through. Because it was originally a military system, the builders of the ARPANET believed that at some point in the future, someone would attempt to take out one or more parts of the network. The network software and hardware were designed to get around this.

The combination of the strong interconnectedness of the network and the extra routes to withstand attacks meant that Morris's one-in-seven rule allowed a positive feedback loop to develop. We came across these in the chapter 4, "Climate Catastrophe"—it's the reason for the squeal of sound when you put a microphone too near a speaker. Imagine you had a mechanical hand turning the control that made the hand move. As the hand pushed the control, it would increase the power, strengthening the push, which would move the control more, strengthening the push even more, and so on.

This same kind of effect was happening with the computers on the ARPANET as Robert Morris's worm took hold. The more the worm was installed on a computer, the more it tried to pass itself on to other computers—and the more copies of the worm were installed on the computer. Before long, hundreds and then thousands of computers were running more and more copies of the worm. And each copy that ran slowed the computer down until the machine ground to a halt.

Just how dramatic this effect was can be judged by a time-based report on the status of the first computer to be infected, a DEC

VAX minicomputer at the University of Utah. It was infected at 8:49 p.m. on November 2 and twenty minutes later began to attack other computers. By 9:21 p.m., the load on the VAX was five times higher than it would normally be—there were already many copies of the worm running. Less than an hour later, there were so many copies running that no other program could be started on the computer.

The operator manually killed all the programs. It then took only twenty minutes for the computer to be reinfected and grind to a halt again. The only way to avoid repeated reinfection was to disconnect it from the network. Ironically, Clifford Stoll, one of the operators responsible for a computer taken over by Morris's worm, rang up the NSA and spoke to Robert Morris Sr. about the problem. Apparently, the older Morris had been aware for years of the flaw in sendmail that the worm used. But at the time of the call, no one knew that it was Morris's son who had started the worm on its hungry path through the network.

The ARPANET survived the unintentional attack, and Morris became the first person to be tried under the Computer Fraud and Abuse Act, receiving a hefty fine, community service, and probation. Morris's worm was, by the standards of modern computer viruses and worms, very simple. It wasn't even intended to cause a problem. But there is a more modern strand of attacks on our now-crucial information and communication infrastructure that is deliberate, intended to cause as much disruption and damage to our society as it can. It goes under the name of cyberterrorism.

Since 9/11 we have all been painfully aware of just how much damage terrorists can do to human life, property, and the liberty of a society to act freely. Although its name contains the word "terrorism," cyberterrorism is at first sight a very different prospect. It may not result in immediate deaths—although at its worst it could result in carnage—but it has a worldwide reach that is impossible for real-world, physical terrorisms. A cyberterrorist attack could target essential sites all around the world at the same time.

The "cyber" part of "cyberterrorism" comes from the word "cybernetics." Taken from the Greek word for steering something (it's the Greek equivalent of the Latin *gubernare*, from which we get the words "govern" and "governor"), this term was coined in the 1940s to cover the field of communication and control theory. At the time it had no explicit linkage with electronics, although the first crude vacuum tube computers had already been built; but it would soon become synonymous with electronic communication and data processing. So cyberterrorism implies acts of terrorism that make use of electronic networks and data, or have our information systems as a target.

Although we tend to think of the threat of cyberattack being primarily at the software level, like the ARPANET worm, because the nature of computer networks makes it easy to spread an attack this way, it is quite possible that a cyberterrorist attack could be at the hardware level. Information technology is always a mix of hardware and software, each being a potential target.

One approach to hitting the hardware is crude and physical. Although the Internet does have redundancy and can find alternative pathways, there are some weak spots. As was demonstrated in New Zealand, destroy a few of the major "pipes" that carry the Internet traffic and it would be at the least vastly degraded. Equally, there are relatively few computers responsible for the addressing systems that allow us to use human language for the location of a computer like www.google.com, rather than the Internet's true addressing mechanism, which uses a string of four numbers, each with three digits. A concerted attack on these addressing computers could cause electronic devastation.

However, a sabotage approach of this kind requires a concerted effort around the world—to be catastrophic, it would need be the most complex terrorist attack ever undertaken. By comparison, there is a simpler hardware approach to cause devastation—and devastation that would spread not just across computer networks, but potentially across all our modern devices. The only constraint here is that it would take a considerable time to set up. But if we are looking at sabotage that might originate from a rogue government or secret state apparatus, this does not seem inconceivable.

Practically every device we use these days, from a sophisticated computer to a TV remote control, contains microchips—tiny wafers of silicon carrying embedded circuits. If saboteurs could gain access to the relatively few companies making these essential components, and make secret modifications to the chips, they could set them up to fail at a set time in the future, or as a response to a particular input.

This possibility has been studied both by academics at Case Western Reserve University in Cleveland, Ohio, and by DARPA. The complexity of many chips is so high that it is entirely possible for there to be circuits in place that are never used or noticed by the manufacturer, but that can later be triggered by timing or an external signal. It is also possible with less technical skill to reduce the lifespan of a chip, causing it to deteriorate and fail quickly.

Although there have been some suggestions for mechanisms to

aid in the detection of such dangers concealed in the hardware, realistically there is very little chance of spotting an extra circuit, even with the sophisticated checking processes used in chip-manufacturing plants. Although the relatively small number of chip manufacturers limits the opportunities for a would-be saboteur, it does mean that if such a cyberterrorist can get in place, the scale of the havoc caused (within a few years, when the terrorist's designs have percolated out through the user base) can be immense.

This is one potential disaster where the only real precaution is good staff vetting in the chip manufacturers, and hope. Yet there is much more to cyberterrorism than the possibility of chips being sabotaged. It's easy to underplay the impact of an act of cyberterrorism, whether software based, the result of physical attacks, or most likely a combination of the two. You may think, yes, it might be irritating to lose the Internet for a few weeks. Some businesses would go to the wall. The rest of us would suffer inconvenience. But we managed without the Internet before it came along, right? This picture is wrong on a number of levels.

First, we are increasingly dependent on the Internet—and that is far from being the only network at risk. We saw in the banking crisis of 2008–9 just how big an impact a failure of a relatively small portion of the banking system could have on the world's economies. Yet almost all banking now depends on electronic networks. If the banking networks were brought down, not only would the banks fail to operate effectively, but commerce as a whole would find it difficult to function.

There are worse possibilities still. All our major utilities—electricity, natural gas, water—rely on computer networks and electrical power to operate. If an attack could take out the power and the control networks, we could see the rapid collapse of the utilities we all depend on for day-to-day living. That electrical power is needed to pump gasoline and diesel fuel too. That means no distribution of goods. Empty supermarkets, with no way to sell items, even if they had the produce on the shelves. It has been said that our modern society is only a few days away from chaos if we lose electrical power. We no longer have the capability to scrape by with what's available around us. Without a support network, towns and cities can't function.

We can see a smaller-scale version of the potential for this kind of breakdown in the blackout of 2003. As we have seen in our discussion of climate change (chapter 4), this affected a sizable part of Canada and the northeastern United States, leaving 50 million people without power. The result was not only loss of electricity, but disrupted water supplies for millions who rely on pumped water, chaotic disruption of transport, massive losses for business, restrictions on medical capacity, limited communications, and looting. This was survivable. But what if the whole continent lost power for weeks?

Some people ignore the threat from direct physical action to disrupt networks because, while it's easy to imagine someone sitting at a computer terminal in a distant darkened room hacking into distant computer systems and wreaking havoc, it's harder to get into the mind-set of someone who will blow other people up—or even kill himself—in order to disrupt society. Particularly American society. Yet all our experience with terrorism to date is that this is just what groups like al Qaeda are prepared to do. It's how they think. So we must not ignore the possibility that cyber-attacks will come in this way.

There have been small-scale attempts using both approaches. For example, in Australia in 2000, Vitek Boden accessed a utility company's computerized control software using a laptop and a two-way radio to intercept the communications between the computers that formed the management system for Maroochy Shire Council's sewage services. He caused the system to release a quarter of a million gallons of raw sewage into public waterways. He had recently been refused a job with a sewage company.

On the physical side of such system-based terrorism, a plot by the former terrorist group the Irish Republican Army was uncovered in 1996. The intention was to use an array of explosives to disable key points on the electricity and gas networks, causing chaos in London. This was potentially devastating—but was much smaller than the possibilities that are now available for truly large-scale cyberterrorism, particularly if applied to information networks rather than utilities.

It might help the terrorists who would resort to direct physical attack if they read academic papers. Scientists at the Dalian University in Liaoning, China, used computers to model how the U.S. West Coast electricity grid would hold up when different parts of the grid were taken out, whether accidentally or in a determined attack.

Before this research was undertaken, the assumption had been that the obvious subsection of the grid to take out in an attack would be a heavily loaded one. The idea was that the systems that control the grid should instantly transfer the load from the damaged subsection to adjacent subnetworks, which would then become overloaded and drop out, resulting in a cascade of failure that could take out the whole network. Surprisingly, the research showed that this wasn't the best approach to take. Under some conditions it was more effective to disable a lightly loaded subnetwork first, if your intention was to cripple as much of the grid as possible. Details of the conditions required for such a collapse have been passed on to the operators of the grid and the Department of Homeland Security.

It has been pointed out, though, that such subtlety is not only beyond most terrorist groups, but unnecessary. Ian Fells from Newcastle University in England has bluntly remarked that "a determined attacker would not fool around with the electricity inputs or whatever—they need only a bunch of guys with some Semtex to blow up the grid lines near a power station."

Although there has been no major attack using purely electronic means, we have seen enough smaller-scale examples to know what is possible. In 1997 the National Security Agency led a dummy attack on a range of networks that would enable it to produce denial-of-service attacks. (A denial-of-service attack is where a server is bombarded with requests and grinds to a halt, rather like the impact of Roger Morris's ARPANET worm, but in this case undertaken deliberately.) The aim was to bring down telephone networks and to block the ability to use e-mail. All this was achieved using tools readily available at the time on the Internet.

It's ironic that the same hackers who undertook the attack on behalf of the NSA pretended to be working for the North Koreans, as in July 2009 we saw a real denial-of-service attack on public computer systems at the White House, the Department of Defense, and the New York Stock Exchange that appears to have originated in North Korea. The only externally obvious result was the disappearance of a number of official Web sites, including the

White House site; but one expert described this as a "massive outage," and even if the actual result was reasonably cosmetic, the potential for damage had this been carried out on a wider scale was considerable.

To make matters more worrying, it is often true that the control systems for our essential utilities, called supervisory control and data acquisition (SCADA) systems, do not have as good a protection from outside interference as do Web sites and other more visible objects on the network. This is partly because those involved aren't always aware that the external connections exist, but also because the level of real-time communication required for these control and reporting systems sometimes means that the more heavy-duty security protocols and devices used by secure Web sites aren't appropriate, because they're just too slow.

There is reasonable evidence that what we hear of these cyberspace attacks is just the tip of an iceberg of threats battering the firewalls and security systems of companies large and small—but most worryingly, also launched against the military and those responsible for key aspects of the infrastructure, such as the power companies. Large companies will get low-level attempts to get into their systems hundreds of times a day, but the power companies experience a serious, heavy-duty attack through the network around once a month. Month after month. Many of them, it has been suggested, are funded and organized by Middle Eastern organizations—or even governments. Occasionally something must give.

But it is not these individual assaults on electronic security that provide the real worry about the impact of a cyberattack. The real concern for those attempting to foil cyberterrorism is that there won't be just a single stand-alone attempt, but a coordinated assault on many of the computers controlling our infrastructure, resulting in countrywide or even worldwide chaos.

We normally think of such attacks coming in through the Internet, routing their way from a distant country and carefully insinuating their electronic tentacles into supposedly secure systems. Although not infallible, most companies and organizations have software (and sometimes hardware) called firewalls, which are designed to reduce the possibility of an external influence on their systems by trapping all incoming messages that aren't internally verified. Yet the Internet isn't the only way of getting into a key system.

Often organizations intentionally open up their systems to the outside world for convenience, in order to be able to use them more effectively. And nothing has caught on more quickly than the ultimate convenience of wireless. Once upon a time, when you paid your tab in a restaurant you had to either give your credit card to the waiter (and hope he wasn't going into a back room and ordering thousands of dollars' worth of goods from Asia with it), or trudge over to the till to make use of a tethered machine to punch in your PIN. Now the device comes to your table. It's wireless.

Many of us have the same kind of convenience at home or in the office. A wireless network means we can connect to the Internet or get data off a static PC from anywhere in the house, using a laptop in bed, or an iPhone in front of the TV. It's great. After all, not many of our homes have network sockets in every room—and even if they did, we would be tripping over wires all the time. But that convenience comes with a price.

It's a sobering lesson to wander around suburban streets with an iPhone or a similar device that latches onto wireless networks as it discovers them. Yes, some are password protected, but many are open. You can just jump in and make use of them. If all you want to do is piggyback on someone else's wireless connection to access the Internet it's not exactly the crime of the century, but the ease of getting onto wireless links does present a danger. And in the early days of such technology, there was limited security.

A particular source of concern a few years ago was the wireless systems used at airports for curbside check-in and to allow airline agents to get access to all the key computer systems as they combed queues for important passengers. Airlines provide a particularly attractive target for cyberterrorists. Though 9/11 and more recent bomb attempts like the December 26, 2009, Detroit incident were conventional plots, airlines operate complex electronic networks worldwide, making them susceptible to cyberattack.

For instance, American Airlines—inevitably one of the terrorists' favorite targets because of its prominence and name—had wireless systems at 250 airports across the United States that in the early years of the implementation weren't encrypted. So anyone monitoring the signal could follow the keystrokes as agents logged on to various systems, making passwords and procedures open to any electronic eavesdropper.

This wasn't just opening up the commercial systems to dangers of manipulation, causing practical problems for airline operations. It also meant that terrorists intending to hijack a plane could find out key information—who were the sky marshals on board, for instance. It could even could allow the terrorists to manipulate the check-in data to make an unexpected bag "disappear" or to confuse the tally of passenger numbers to conceal an extra flier who hadn't passed through security.

Since these threats were uncovered, the airlines have taken measures to ensure that their wireless communications are less easy to monitor. But there is still a lot of unprotected wireless traffic out there, and the chances are that some of it will be in areas where allowing open access creates a security risk.

The outcome the public fears most in this arena is that a cyberterrorist could hack into military computers in charge of nuclear weapons and start World War III. This was the scenario of the 1983 movie *WarGames*, where a teenager gets access to a Defense Department computer, believing it to be a game, and brings the world to the brink of war. Such a simplistic attack is not going to succeed—ever since the formation of MILnet, most low-security military computers have been off the Internet, and the high-security ones never were connected. The military has no need to use commercial networks for this kind of activity.

In principle, it is possible to make a computer that is unhackable, simply by not connecting it to anything else. This is likely to be the case for the most dangerous systems, though some military computers could be open to a more direct cyberattack, for instance by tapping into the optical cables used to connect them. In practice, though, cyberattacks causing a nuclear holocaust can be left with Hollywood—the hijacking of the systems controlling our businesses and utilities could still cause chaos with much less effort and risk.

We aren't going to stop communicating, or making use of information technology. It has become an essential part of everyday life. But those who are responsible for the key systems that run our infrastructure and defense need to be constantly aware of the threats we face. Communication is, after all, one of the essential behaviors of human beings. But there are other aspects of humanity that could come under threat. Perhaps the most subtle peril the human race faces is that we could cease to be human beings at all, as the ever increasing pace of biotechnology enables something better to come along and replace us.

CHAPTER EIGHT
NO LONGER HUMAN

||||||||||||||||||||||||||||||||||||||||||||||||||

> *The human race began to mellow then.*
> *Because of fire their shivering forms no longer*
> *could bear the cold beneath the covering sky.*
> —Lucretius (ca. 96–ca. 55 BC), *On the Nature of Things,*
> trans. Anthony Esolen (1995)

The idea that human beings are driven to go beyond nature is nothing new, as the quote from a Roman natural philosopher at the start of this chapter shows. In my book *Upgrade Me,* I suggest that the urge to enhance ourselves is part of what makes us human. Our technologies, and the changes that we have made to our bodies, both physically and chemically, have gradually given us capabilities that go far beyond anything that early human beings could do.

It all started with the ability to think outside the here and now, to consider what might be. Once we were asking "What if?" it was inevitable that we would begin to look for ways to enhance the basic capabilities of the human body. There were at least five separate drivers for making these changes. Just like us, the early humans wanted to become more attractive to the opposite sex, to put off death, to enhance their strength to defend themselves more effectively against physical threats, to make better use of their re-markable brains, and to repair damage to their bodies.

Just consider how much progress we have made, how much we have modified ourselves and our environments, in each of these categories. If we take the single category of putting off death, we need to include everything from preventive medicine to the simple but powerful act of cooking food. More recently we've seen a huge amount of effort put into science that may enable us to extend the human life span, whether by preventing the action of killers like cancer or by making changes to the fundamental aspects of our biology that promote aging.

Along the way, quite unintentionally, we have subtly changed what it is to be human. Biologists will tell you that we are still the same animal that we were 100,000 years ago. Homo sapiens as a species has not evolved in a biological sense. But in terms of function and capability, in terms of the way our bodies work with the various enhancements we provide, we are already human version 2.0, a whole new creature.

There is nothing wrong with the urge to upgrade. It really isn't exaggerating to say that without it we wouldn't be human beings—so it's meaningless to simply try to discard it. Some argue that we need to say, "Stop, enough, we don't want any more development of the human being." Such commentators really don't understand people. You might as well say, "Stop now, we don't want there to be any more breathing, it's bad for the environment." Yet there is the chance that our pressing demand for more and more enhancement will in the end prove terminally destructive to the human race.

A relatively low-risk possibility is that our attempts to improve on nature will result in a medical catastrophe. It is possible to imagine a scenario where, for example, a hugely popular cosmetic caused a worldwide breakdown in the human immune system, or a common, everyday drug proved to have massively damaging long-term side effects. But it's unlikely that any single product will ever have large enough market penetration to have a truly devastating effect on the planetary population, and products that do have a vast reach always have a wide range of testing incorporated into their design.

A more subtle variant on the medical catastrophe would be if a universal, or near-universal, product rendered us particularly vulnerable to a pandemic from an existing virus or other illness that in itself could not cause mass death, but that was transformed into something more dangerous by the product.

We can never be 100 percent sure of the impact of a medical treatment or of a product we consume. Witness the tragedy that resulted from an unexpected side effect of the drug thalidomide, which was used in the late 1950s for a number of purposes, notably suppressing morning sickness. Between ten thousand and twenty thousand children would be born with birth defects as a result of their mothers taking thalidomide.

It's a horrible story—but it also reflects the relative difficulty of inadvertently causing a disaster on a scale that would wipe out a major portion of the human race. The medication would have to be taken by a huge population across the world, but such of widespread distribution of drugs is rare with anything less tested and safe than aspirin.

Of course, not all human upgrading is harmless. There is one technological enhancement to our bodily capabilities that already kills forty-five thousand people each year in the United States and more than a million people yearly worldwide. It's hard to imagine that we can tolerate a product that has proved so deadly, but the fact is, we do. We can't get by without it. It has become such an integral part of our nature that we couldn't contemplate managing without this upgrade, even though it is responsible for so high a mortality rate—and, what's more, it damages the environment at the same time.

This rampant killer is the car (or, more precisely, road vehicles). Yet even at this level of deaths, we aren't looking at an enhancement that is in any danger of wiping out humanity (unless you consider climate change). But some believe that we are not far from being able to upgrade human beings so much that normal, unmodified humans will eventually become extinct.

This is the possibility that future gazer Ray Kurzweil refers to as the Singularity, based on an idea first conceived in the 1980s by science-fiction writer Vernor Vinge. Vinge predicted that "within 30 years, we will have the technological means to create superhuman intelligence. Shortly thereafter, the human era will end." His prediction, which should have been coming to fruition around now, overestimated the speed of technological development; Kurzweil now suggests that the Singularity is likely to be reached around 2040. Yet the threat remains the same.

In this vision of the future, human beings become merged with computing technology to form a new type of life, a unique hybrid species with vastly enhanced thinking capabilities that would render ordinary human beings obsolete. This new intelligence may continue into the future with a biological-electronic hybrid brain, but is more likely to discard the biological aspect entirely as the capabilities of electronics continue to be exponentially enhanced.

With, perhaps, an unemotional view of the nature of life and what it is to be human, it seems likely that any human-computer hybrid would not be content with the capabilities of a fleshy body. It would begin to replace the less effective physical aspects of the human frame. As enhancement technology improved, more and more of our biological functions would probably be surpassed by a constructed alternative, or a biological-mechanical hybrid.

The idea of discarding the biological entirely was suggested as the most likely outcome by a trio working at the new technology labs of the British telecommunications company BT back in 1995, looking speculatively at the "future evolution of man." Ian Pearson, Chris Winter, and Peter Cochrane imagined that by around 2015, computing and robotics would be so advanced that Homo sapiens would be overtaken in all abilities by a robot species they dubbed Robotus primus.

At the same time they imagined human beings becoming sufficiently enhanced with electronic implants to be considered a new species themselves—Homo cyberneticus. But this first cyborg, they felt, was a transitional move to Homo hybridus, which would have biological enhancements as well as the cybernetic. (Back in 1995 it wasn't obvious that genetic engineering would race ahead of the interface between man and machine.) This hybrid, too, was expected to be relatively short-lived. As the capabilities of electronics continued to grow, the parts of the hybrid where normal biological processes were better than the electronic would become fewer and fewer. Between 2100 and 2150, Pearson, Winter, and Cochrane imagined, we would abandon flesh and become purely electronic—Homo machinus.

This new species would be quite different from anything we know. In the BT trio's words, the citizens of the future would see a creature that was "vastly more intelligent and has access to whatever physical capability is required. It can travel at the speed of light, exist in many places at once, and would be essentially immortal. It would coexist with Robotus primus, but we could expect that the two would closely interact and may quickly converge."

They also envisaged that many humans would reject enhancement, initially living in parallel with the increasingly enhanced forms, as perhaps Neanderthal man did with the early Homo sapiens. Sadly, the BT visionaries predicted that peaceful coexistence would not be sustained for long, and the remnants of Homo sapiens (or Homo ludditus, as they condescendingly christened our descendants) would soon be rendered extinct by their creations.

It is interesting to make that comparison to Neanderthal man, a second human species that coexisted with Homo sapiens for thousands of years. Recent research has shown Neanderthal man in a new light. Traditionally portayed as apelike, hairy, and unintelligent, it now seems likely that Neanderthals were fair skinned, looked much more like existing humans than like apes, and had language and a fair degree of intelligence. It was long thought that Neanderthals were wiped out by the deadly, competitive Homo sapiens, but it now seems that Neanderthals were simply less flexible, more stuck in their ways, and so didn't make the changes in lifestyle necessary to survive the last ice age. Perhaps there are lessons to be learned for any future members of Homo sapiens confronting a new hybrid human race. But it seems unlikely this will be an issue for many of us alive today.

The BT time frame (particularly the development of a superior robot species by 2015) now seems as far out of touch with reality as Vernor Vinge's, but there is a certain amount of logic to the progression they envisaged. Although such a cyborg future human would not have to take on the hive mind of *Star Trek*'s frightening Borg aliens, it is possible that they would share either the Borgs' urge to assimilate other creatures, or at the very least a tendency to oust nonmodified humans first from positions of power and skill, and then from existence. This would not require any malice on the part of the new race. It would be sufficient that they were functionally superior and competing for the Earth's limited resources. Over time, they would come to dominate.

However, before we all panic about our impending replacement by our cyborg superhuman creations, it is worth noting a few lessons from history—all too often, it seems those who predict this kind of future are relying entirely on science fiction for their guide to the future, whereas the past can sometimes be more usefully informative. Science fiction is not an oracle that sees into the future—the key word here is "fiction."

The first consideration that throws doubt on a future when true humans are ousted by electronic hybrids is that just because we gain added capabilities doesn't mean that we will lose control of the upgrade, so that upgraded individuals become something other than human. Our existing upgrades in the form of a car and a computer, for example, have vastly more physical power and speed of computation, respectively, than is available to a basic human being, but we have ensured that they remain subservient to the human will. There is no reason to be sure that with any future enhancement, we will suddenly lose this upper hand.

The other problem with Kurzweil's vision is that it depends on a risky approach to predicting the future of technology. The assumption is that there will be intelligent, conscious technology that will engineer this replacement of the basic human in a relatively short time (by around 2040 on Kurzweil's predictions), because the growth in capabilities of information technology has to date been exponential, getting more and more powerful at an explosive rate.

We have often seen technology develop like this—but it also has the habit of suddenly hitting a plateau or even reversing the trend. This is something that those predicting the Singularity seem to have missed. At the moment, for example, artificial intelligence and robotic development are not seeing anything like the speed of development that we are witnessing in basic information technology. Even if true thinking robots were getting closer as quickly as processor speeds were being enhanced, it isn't a safe bet to assume that developments would continue at the same hectic pace.

Science writer Damien Broderick has pointed out, as a model for the way computing technology and robotics will develop, the way that human speed of travel has changed. For millions of years, humanity moved around by foot, with no other choice. A few thousand years ago we speeded up a little by using horses and other animals. It was just two hundred years ago that we added the steam engine to our kit, and since then we have seen a rapidly accelerating speed curve, with cars, planes, jets, and rockets.

This is a seductive but flawed argument. We achieved rocket flight into space back in 1957, it's true, but if things were continuing to speed up exponentially, we should have come up with several quicker means of transport since then. We haven't. Admittedly there have been tweaks and technical improvements in space flight, but nothing much quicker has emerged. The developments in transport have decayed since 1957. More important, this picture of ever-increasing speeds does not even compare like with like.

It is only really meaningful to make a comparison of the transport available to ordinary citizens, who have never had access to rocket travel. For us, the last big increase in speed was the introduction of supersonic commercial airliners with Concorde in the early 1970s. At around 2,000 kilometers per hour (1,300 miles per hour), this was a big step forward from conventional jets, more than doubling their speed. Yet since, not only has there been no enhancement, but with Concorde's retirement we have fallen back to the speeds available thirty or forty years ago. This could be just as good a model of the way intelligent technology could develop as visions of computer capabilities that continue to shoot up an exponential curve.

Yes, there is a risk from our rampant drive to self-enhance, but it is much smaller than many of the dangers we face as a human race. Those who predict the Singularity have a false picture of the nature of humanity. They see us as the biologists do, as being just human 1.0. But we have already come this far, and from the evidence to date, the enhancements they envisage wiping out humanity are not likely to be so extreme, but will simply continue the incremental modification and growth in capability of that most remarkable species that is Homo sapiens.

Although we have used enhancements for thousands of years, the more dire possibilities for our destruction by enhancement are attributed to technologies we have yet to fully develop. The possibility that normal, unenhanced humans will become outdated, made obsolete by cyborgs or artificial intelligence, is still remote. As we look into the future, this is not the only threat we face, either from technology or from the natural world.

CHAPTER NINE
FUTURE FEARS AND NATURAL PITFALLS

||||||||||||||||||||||||||||||||||||||||||||||||||||||||||||||||||||||||||||||||||||||||||||||||||

Predictions can be very difficult—especially about the future.
— Niels Bohr (1885–1962), quoted in H. Rosovsky,
The Universe: An Owner's Manual (1991)

When we consider the disasters that could befall us in the future over and above the areas we have already explored, there are two possible culprits: natural disaster and the outcomes of our own interaction with the planet. Sometimes the two are hard to separate. Although it's unlikely we would be responsible for an asteroid collision with the Earth, it is entirely possible that our interference could trigger seismic activity. But whether or not we do mess things up, we know from history that there are a whole range of natural phenomena just waiting to give life on Earth a pounding in the future.

The first distinct possibility for widespread human destruction has a clear precursor in a mass extermination that has been subject to some subtle crime-scene investigation, an ancient genocide solved thanks to our knowledge of the element iridium. The concentration of iridium in meteorites is considerably higher than it is in rocks on the Earth, as most of the Earth's iridium is in the planet's molten core. One class of meteorite particularly, called chondritic (meaning it has a granular structure), still has the original levels of iridium that were present when the solar system was formed.

In 1980, a team led by physicist Luis Alvarez was investigating the layer of sedimentary clay that was laid down around 65 million years ago, a time of particular interest because this so-called K/T boundary between the Cretaceous and Tertiary periods marks the point at which the majority of dinosaurs became extinct. ("Cretaceous" is represented by K because C was already allocated. "Tertiary" is now a largely disused term, and it should more properly be the K/Pg or Cretaceous-Paleogene event.) This layer contains considerably more iridium than would normally be expected, suggesting that there may have been a large meteor or asteroid strike on the Earth at this time.

There is so much iridium present that the asteroid would have to have been over six miles across—large enough to devastate global weather patterns, bringing about changes in climate that could have wiped out the dinosaurs. This was neither the first nor the last strike by an asteroid or comet that the Earth has faced. It is a constant threat, and one that we are only now starting to wake up to.

The most recent sizable impact was the Tunguska event in Siberia in 1908. Although some still try to cloak this occurrence in mystery because there is no impact crater, it is generally accepted in the scientific community that an extraterrestrial body, probably a comet, exploded above the Tunguska plain, flattening trees for miles around and leaving a shattered terrain. If this impact had taken place in a city rather than in the wilderness, it would have had an effect similar to that of a nuclear explosion in terms of blast, if not of radioactive fallout.

There's a one in five hundred chance of an event on the scale of Tunguska or bigger happening each year—not hugely probable, but distinctly possible. It has been estimated that the odds are only one in four that the strike would kill anyone, and one in seventeen that it would result in mass-scale mortality with ten thousand or more dying. This surprisingly low risk reflects how much of the planet's surface is either under water or not highly populated. Even so, the risk is there, and if a large enough body did hit an ocean, it could generate a vast tsunami, which could cause almost as much devastation as would a direct impact.

Although there are asteroid surveys under way that should give us a few days' warning of an incoming piece of space debris on a collision course, these are limited in scope, both in their coverage of the skies and when they can operate. There are plans for future observatories that would give us much better coverage, but for the moment it is entirely possible that an asteroid strike will occur without warning.

The nightmare scenario for Armageddon is that such a strike will be interpreted as a human attack and will result in retaliatory launches of missiles leading to a nuclear holocaust before it can be established what caused the first impact.

Some movies would have us believe that collisions with asteroids don't really present a problem. All we need to do, they suggest, is to blast any incoming asteroids to dust with nuclear weapons. This was also the approach favored by H-bomb enthusiast Edward

Teller. However, there are some real problems any would-be aster-oid blaster would face. First, we'd need plenty of warning. At the moment it is only in a few rare cases that we would have the years of preparation needed to construct and fire off the appropriate device. It's not enough to take any old nuclear warhead and fire it into space, as neither the bomb nor the delivery rocket was de-signed for this task.

To make matters worse, many scientists think that a nuclear bomb would just shatter an asteroid into an array of chunks, still big enough to cause devastation, but now spread over a wider area of Earth's surface, rather than arriving at a single point of impact. There's also the minor matter that there is a treaty preventing the use of nuclear weapons in space—though this could probably be got around if the survival of the world depended on it.

In practice, our best hope is not deflection, but to build a better warning network so that we can give those who are likely to be anywhere near the impact zone days of warning to evacuate. This still won't help us if we were faced with the kind of impact that wiped out the dinosaurs. That would have planetwide effects. But for any likely impact in the next few hundred years, a good warn-ing system should deliver the results. Until then—we'd better all hold our breaths.

One of the inevitable outcomes of a collision with a body from outer space would be seismic disturbance, but we don't need a space invader to get the Earth moving beneath our feet. The natu-ral processes of the planet are perfectly capable of producing dev-astating results. Although to the casual observer the Earth's solid surface is immobile and rigid, seen on a large scale with a longer time frame than year-by-year human observation, it is anything but.

The outer skin of the Earth floats on the hot, fluid material beneath. You've probably seen film from the International Space Station, where globules of liquid float in space. Imagine an immense drop of such a liquid that forms a crust on the outside. The crust is split into a number of segments, and convection flows in the liquid are constantly trying to move those segments. Where these plates are forced against each other, or one segment of crust is pushed under another, vast forces come into play. This is the how the Earth is built. And the outcome of such a collision of plates is an earthquake.

Earthquakes can have a double impact on human beings. First there is the direct damage, most often caused by falling buildings. But also an earthquake in the sea can cause a tsunami, which we will return to a little later. Many earthquakes have relatively little impact—the vast majority are so low in power that they are not noticed by casual observers—but a few have catastrophic effects. At least two earthquakes in China have killed hundreds of thousands of people each, while the earthquake that struck Haiti on January 12, 2010, is thought to have killed over 200,000 people.

Although earthquakes have been recorded since ancient times, it is only very recently that we have begun to understand what is happening. At one time they would have been considered a ground-based relative of thunder and blamed on the action of gods. It was only in the eighteenth century that any real attempt was made to explain these strange forces, with an idea dreamed up by a man who lived in one of the countries least affected by earthquakes, England.

In the 1760s, the astronomer and geologist John Michell, whom we've already met dreaming up the black hole, put forward the idea that earthquakes were caused by underground steam. The timing of this deduction was not accidental. The power of steam was just beginning to be realized in Britain's Industrial Revolution, with steam-pumping engines replacing manual pumps in mines. Steam was capable of delivering a power that went far beyond any human or animal capability. Michell believed that build-ups of steam underground were responsible for the tremors and shifts in the Earth that occur in earthquakes.

He might have been wrong about the cause, but Michell did work out the way that shock waves pass through the Earth's surface to produce quaking. Others would blame earthquakes on the influence of the Moon, but it was only with the twentieth-century concept of continental drift, later developed into plate tectonics, that a convincing explanation for the forces behind the earthquake, based on those vast shifting plates colliding and passing over each other, was developed.

If we are to avoid significant devastation caused by earthquakes we need to know where they are likely to take place, to build structures in such areas that can withstand earthquakes, and to be able to forecast an earthquake's arrival to be able to evacuate as many people as possible from the danger zone.

It's fair to say that we are getting better at this. We have good information on areas that are at risk, and of the probability of future quakes in the general sense. And in some earthquake-prone areas, like Japan, a huge amount of effort has been put into constructing buildings that can withstand high levels of seismic activity without collapsing and crushing those within. But in poorer countries there are still many buildings that cannot stand up to expected shock levels, let alone the less frequent quake of really major proportions.

Although we have to continue building more sensibly in earthquake-prone regions, we still need warning of a quake coming. Like weather forecasting, this is never likely to be possible beyond a certain time frame—and for the same reason. The Earth's weather system is chaotic in the mathematical sense. Very small changes in initial conditions can make huge differences a few days down the line. In the last few years, meteorologists have made giant steps forward in the accuracy of short-term prediction, but we will never be able to forecast accurately several weeks out.

Similarly, although it may be possible in time to predict an earthquake with reasonable accuracy a few hours before it occurs, we will never have an ideal warning, weeks in advance. Rather like heart attacks (which also have chaotic features), some earthquakes do have warnings in the form of preshocks, but others don't. Predictions have been made already that have saved lives; but predicted earthquakes have often not come about, and if forecasts are repeatedly inaccurate they will soon be ignored. A lot of work is going into earthquake prediction, but the jury is still out on what's possible.

The other significant geological threat to human life comes from volcanoes. These are formed when magma, molten rock, rises under pressure and bursts through the crust. An eruption can be a relatively gradual process, or an explosive one, usually becoming explosive when an underground volume of water is flash-boiled by the magma, blowing out a section of the Earth's surface (surprisingly similar to Michell's mechanism for earthquakes).

Volcanoes are always with us—there are estimated to be around 1,500 volcanoes at the moment that present a potential danger. Often it is possible to escape an eruption where the only hazard is a lava flow, but communities can suddenly be engulfed by tons of ash, as happened at Pompeii and Herculaneum in Italy when Mount Vesuvius erupted in AD 79, leaving amazingly well-preserved remains.

Even relatively small eruptions in the present day can cause death and disruption. The Mount St. Helens eruption in Washington State that took place in 1980 killed fifty-seven people, though there was enough warning to give everyone the chance to get to safety. At the other extreme, the Krakatoa eruption in 1883 killed around thirty-five thousand people, and the less well-known Indonesian eruption on Sumbawa in 1815 had a death toll that reached fifty thousand.

Unlike earthquakes, because a volcano is a known, specific target to watch, it is possible to get much better warnings of potential eruptions. A whole gamut of monitoring technology is now used in high-risk locations. Small explosions are used to generate shock waves, which are tracked to map the progress of the magma within the Earth, while variations in both electrical conductivity and magnetic fields can show the increasing nearness of an event, as can subtle bulging movements in the outer layers of the volcano itself, as if the mountain were a cartoon figure, puffing up ready to vent its wrath.

The newest possibility for monitoring volcanoes comes from scientists in Japan, a country with more than its fair share of volcanic activity, who are using short-lived particles called muons to keep an eye on what's going on under the ground. Muons are produced in the atmosphere when cosmic rays—high-energy particles from the depths of space—crash into the upper atmosphere.

Muons have a very short lifetime and few should make it to ground level, but they travel so fast that special relativity plays its part. Because of their relative speed, the time the muons experience is significantly slowed down, by about a factor of five. The result is enough muons penetrating volcanoes to be able to use them to monitor the amount of magma inside a volcano, as the percentage of muons that gets through depends on the mass of the material inside.

But for all the disruption that a regular volcano can cause, it is trivial compared to a supervolcano. Supervolcanoes are on a totally different scale—and one of them could erupt at any time in Wyoming's best-known natural attraction, the Yellowstone National Park.

The remains of this supervolcano were discovered by Bob Christiansen of the U.S. Geological Survey in the 1960s, and they provide the classic example of something so big it's difficult to spot. All the geothermal activity in Yellowstone, producing those famous geysers, implies there's a volcano around somewhere. But where, Christiansen wondered, was the crater? It was only by coincidence that he saw some NASA high-altitude photographs of the park and realized that the caldera of the Yellowstone volcano—the pit left after its most recent volcanic explosion—was so big that it hadn't been spotted: fifty-five by seventy kilometers (thirty-five by forty-five miles) across.

The scale of eruption of a supervolcano like Yellowstone—and around half a dozen others have since been found around the world—is truly tremendous. If it were to explode again, it would devastate hundreds of miles around it. In its most recent truly large eruption, nearly 640,000 years ago, the Yellowstone volcano produced enough ash to cover the whole of California six meters (twenty feet) deep. In practice the ash spread over what would now be nineteen states, the vast majority of the United States west of the Mississippi river. Leaving aside the immediate deadly impact, if such an eruption were to happen today it would leave those nineteen states with a vast amount of debris to clear before they could even vaguely return to normal.

Just as the effects of Krakatoa were felt around the world, so the much bigger supervolcano eruptions would not be limited in impact to their immediate surroundings, however large. It's 630,000 years since the Yellowstone supervolcano erupted, but there have been more recent supervolcano eruptions, including Toba in Sumatra some 74,000 years ago, where the distribution of ash in the atmosphere cut back sunlight to such an extent that there was, in effect, a six-year-long winter. Such a change in the Earth's climate would totally devastate agriculture, and billions could die.

Luckily for us, supervolcano eruptions are not common. If they were, the chances are that life on Earth would never have reached the complexity it has. However, we can't be totally complacent. It is impossible to predict when a supervolcano will next erupt, but we do know the rough frequency with which their eruptions occur. The Yellowstone supervolcano seems to blow its top around every 600,000 years. The next eruption is due any day now—though geological "any day now" can easily be give or take a few tens of thousands of years.

The possibility of a supervolcano eruption is terrifying, but the odds are fairly long against it happening in our lifetimes. Besides, there's nothing we can do to prevent it, and little we can do to prepare for an event of this scale. Each year, though, there are other devastating natural events, phenomena of wind and wave, that we should be better prepared for: hurricanes and tsunamis.

There is a certain amount of confusion over just what a hurricane is. It is a powerful storm that arises out at sea, forming huge, slowly spinning spirals that are often as much as 30 or 50 kilometers (twenty or thirty miles) across, and have been known to be three hundred miles wide. The spin is usually counter-clockwise in the Northern Hemisphere and clockwise in the Southern. Hurricanes can drift for days or even weeks before finally dispersing, usually after they make landfall.

Although hurricanes are very obvious on weather-satellite images, their paths of destruction are hard to predict, as they can suddenly veer, or even double back on themselves. Part of the confusion that surrounds them arises from the way that the same phenomenon is given different names in different parts of the world. Though they're hurricanes in the North Atlantic, Caribbean, and parts of the Pacific, they are also known as cyclones around the Indian Ocean and as typhoons in the rest of the Pacific and the China Sea. There is no distinction—they're all the same phenomenon.

Hurricanes can be massive killers—in fact, they kill more people on a regular basis than any other natural phenomenon. In 2005, Hurricane Katrina brought home just how much devastation could be produced even in a highly developed country like the United States, when that storm left behind such a trail of misery and destruction in New Orleans, laying waste around 80 percent of the city, leaving 1,500 dead and many, many thousands homeless—

and this was with enough warning to evacuate most of the residents.

But for the true power to devastate we have to look back a little further, to 1970, when in November a cyclone hit what is now Bangladesh and was then East Pakistan. Blasted by 150-mile-an-hour winds, the vast, low-lying coastal areas were inundated with a surge of water six meters (twenty feet) high, covering over 20 percent of the country's land area during the night, catching many people unawares. With no way to spread a warning and little opportunity to escape, nearly half a million people perished.

Hurricanes are easy enough to spot, especially with modern weather radar and satellite monitoring. The difficulty is being sure of their route, getting a warning out in time, and enabling people to reach safety. Much of the time a hurricane's path is predictable, but because of the sudden swerves one can make, constant monitoring and updates are necessary. In areas like Bangladesh, communications have been improved by the use of cell phones and other modern technology, while concrete shelters have now been built, recognizing the impracticality of mass evacuation on the scale needed to survive a major cyclone—and these have helped save lives in more recent storms.

Often, much of the devastation from a hurricane is caused by the storm surge, where the high-speed winds force the sea to rise and breach the usual defenses. Another, different kind of sea surge is a tsunami or tidal wave. This is usually triggered by underwater seismic activity—an earthquake, or eruption—or it could even result from a meteorite impact on the sea. Unlike waves whipped up by the wind, the tsunami is a solitary wave front, a vast wall of water that moves ahead implacably and destroys buildings like toys.

December 26, 2004, saw the worst tsunami in modern times take shape in Southeast Asia. Around 200,000 people were killed in a sweep of countries bordering on the Indian Ocean. On a smaller scale, but still with devastating effects for the inhabitants, another tsunami struck Samoa at the end of September 2009. Some countries have done what they can to minimize the impact of tsunamis. Japan, for example, has built special coastal walls, while other at-risk areas have planted trees on the shoreline to try to reduce the impact of a solitary wave.

There are also warning systems in place in a few countries— Japan, again, and the United States—that monitor the likely origins of tsunamis and issue alerts to try to get people moved away from the coast in time. There are even a few examples where specific warning signs have been spotted on the beach before a tsunami hit, enabling individuals to run to safety. The water may bubble or take on a strange smell, while the seas recede prior to the arrival of the wave. Such auguries may not provide much of a warning, but they're better than nothing. For now, though, in the often very poor areas most at risk from tsunamis, there is still little chance of receiving a good enough warning to safely evacuate the coast.

Such extreme weather and seismological events may produce tragedies, but at least they are understandable as an extreme form of everyday experiences. Even a meteor or comet, despite its extraterrestrial origins, is, in the end, a collision with a lump of rock or ice, as straightforward as a car crash. But space is also home to some more insidious natural threats.

Perhaps the most obvious worry, when we look out into space, is that our Sun will go nova. "Nova" just means "new" in Latin, referring to a time when a nova was seen as a new star (*stella nova*) in the sky. Early astronomers would occasionally be surprised to discover a star where none had been seen before. Often these got brighter over a period of time before eventually fading away.

We now know that a nova is a star that has exploded, flashing into unparalleled brightness so that it suddenly becomes visible from Earth whereas before it was far too faint to see. In the process it will have totally destroyed any planets around it. If the Sun went nova, the Earth could not survive. Nothing can travel faster than light, so we would not be aware for the first eight minutes after the explosion, the time light takes to reach us from the Sun, but after that, in a matter of moments, all life would be extinguished.

To be precise, if the Sun exploded it would be a supernova, not a nova. This is because the terminology of novas has changed over the years. Although "nova" originally just meant a new star—any star that's suddenly bright—it now applies to a particular way this can happen, when a certain type of a star, a white dwarf, pulls material from a second star that is its binary companion, the two stars orbiting each other like the Earth and the Moon. The new matter, primarily hydrogen, sucked onto the white dwarf from its neighbor forms a thin, high-pressure layer, which undergoes a thermonuclear explosion, like a vast hydrogen bomb.

Only the outer layer is blown away in the explosion, and the nova can then re-form, as the white dwarf sucks more material from the companion star, on a regular basis. This clearly isn't going to happen to the Sun, which isn't a white dwarf and doesn't have a companion star. But no companion is required for the more devastating explosion that is a supernova.

Here, an aging star begins to collapse as the gravitational pull of its mass overwhelms the outward pressure from the nuclear reaction that keeps it alight. As pressure increases, extra nuclear fusion processes that had not previously been possible, such as carbon fusion, may take place. While up to this only a tiny portion of the material in the star has experienced the right conditions to undergo the fusion process, now a good proportion of the material in the star undergoes fusion all at once. The result is a massive stellar explosion that would inevitably take out the whole solar system if the Sun underwent it.

Luckily for us, only certain kinds of stars, at particular points in their life cycle, are able to go supernova. The Sun doesn't fit the bill. To become a supernova it would need either to be much older or much heavier. The Sun is likely to be around for several billion years before it gives us any trouble. But this doesn't mean that we are entirely safe from attack from the depths of space.

This sounds like the plot of a B movie, but it's not a matter of alien invasions (something we'll consider in a moment). Instead, we are at risk from the sinister-sounding gamma ray bursts.

These may be one of the side effects of a star that collapses without going supernova, though we are not certain exactly why they happen, and there are a number of theories battling it out to be accepted as their cause. Everything from collapsing stars to evaporating black holes has been blamed for the production of the bursts. We just know for certain that they're out there and they're very, very dangerous.

A gamma ray burst is an incredibly powerful blast of the most potent rays in the electromagnetic spectrum, which can last anywhere from a fraction of a second to an hour. This doesn't sound too scary, but gamma rays are so energetic that they can cause massive disruption to living things—it's gamma rays that cause most of the damage in nuclear radiation. Simply put, they're killers.

The scale of a gamma ray burst is awesome—in that concentrated blast, the burst will carry as much energy as the Sun is going to give out in its whole lifetime. The good news is that they seem to be rare phenomena. We see only a few hundred a year, and they are so bright that we would expect to see most of the bursts that are occurring throughout the visible universe. This being the case, a burst that threatened the Earth would probably occur only once every few million years.

If we did experience a gamma ray burst close enough to cause damage—which would mean within a few thousand light-years—there would be considerable immediate genetic damage to life on Earth, but more disastrous would be the destruction of the ozone layer. The energy of the gamma rays, slamming into the atmosphere, would cause nitrogen to react with oxygen, forming nitric oxide, which is highly reactive with ozone. Without the protection of the ozone layer, much more ultraviolet light would get through, and it would be this lower-energy but still dangerous light from the Sun that over a number of years would cause the cellular damage that could result in the loss of practically all life on Earth.

Of course, some science-fiction authors would have us believe that dangerous rays are more likely to emanate from the weapons of spaceships belonging to visiting aliens. As weapons of mass destruction go, the ray gun had a surprisingly early conception. To see its first origins, we need to go back to ancient Greece.

Here in Syracuse, on the island of Sicily, Archimedes was born in 287 BC. He is most remembered for his mechanical inventions and for carrying on Euclid's mathematical work. Archimedes certainly had an obsessive enthusiasm for geometry. Plutarch, writing 350 years later, wryly observed that Archimedes' servants had to drag him from his work to get him to the baths to wash him, and when he was there, Archimedes would still be drawing diagrams using the embers of the fires, and even marking out lines on his naked body as he was being washed and anointed.

Archimedes lived in an unsettled time for Greece. The Romans, whom the Greeks had contemptuously dismissed as insignificant barbarians, were sweeping across Greek territories. The once great Hellenic civilization was on the verge of collapse. And Archimedes, for all his genius, ended up in the wrong place at the wrong time. He had designed engines of war that were used to bombard invading ships, but despite these, the Romans seemed unstoppable.

It was 212 BC. With the enemy closing in on Syracuse, Archimedes had the inspiration of using light itself as a weapon. He knew that small, curved mirrors could concentrate the rays of the Sun enough to set kindling alight. This ability to focus energy at a distance seemed an ideal way to attack the Romans' vulnerably flammable wooden ships before they were even in range of the projectile weapons Archimedes had arrayed along the quayside.

Archimedes drew up plans for great curved metal sheets to be fixed in frames on the harbor walls. These dazzling constructions were intended to capture the Sun's rays, focusing them to a point until the undiluted heat of the day became a miniature furnace. But the mirrors were never made. Perhaps the craftsmen, more used to blacksmithing than to precision engineering, found their construction too much of a challenge. Perhaps the stricken city had lost so much to the war effort that it could not find time and money to construct the mirrors. Perhaps even the great Archimedes was laughed at when he claimed it was possible to destroy their Roman enemies without even touching them.

It may have been the mirrors that Archimedes was still working on in his last minutes. According to some legends, he was drawing and redrawing diagrams when one of the invading Roman soldiers found him. Without looking up, Archimedes cursed the interruption: "Do not disturb my diagrams." They are said to be his last words. The soldier who found the seventy-five-year-old man was in no mood to tolerate such disrespect from a member of a defeated nation. Archimedes was slaughtered without compassion.

The concept of destructive energy rays, whether heat or light, resurfaced regularly in fiction from the nineteenth century onward. A typical example was the devastating energy source called "vril," dreamed up by Victorian author Edward Bulwer-Lytton in a book called *The Coming Race.* Vril was a power source that could do anything from drive vehicles to emit a destructive beam that would disintegrate an enemy.

The only way vril would later be remembered would be in the name of a popular British hot drink made from meat extract called Bovril, while Bulwer-Lytton is now best known as the author of the book that inspired cartoon character Snoopy's fictional efforts. Bulwer-Lytton's novel *Paul Clifford* is generally considered to have the worst opening passage of any book in history. It begins, "It was a dark and stormy night; the rain fell in torrents— except at occasional intervals, when it was checked by a violent gust of wind which swept up the streets (for it is London that our scene lies) . . . ," and Bulwer-Lytton's name is now given to a contest for producing the worst opening line for a hypothetical novel.

The real death ray—or at least a beam of light that is capable of killing—emerged accidentally when a pair of Russian scientists were investigating the behavior of the pungent gas ammonia, in 1954. Some thirty-seven years earlier, Einstein had predicted that it would be possible to set off a kind of chain reaction producing light, which he described as stimulated emission.

According to Einstein's theory, an electron in an atom can be pushed into a high-energy state when it is hit by a photon, leaving it like a bucket of water sitting over an open door. Another photon, hitting that electron, would not only be re-emitted itself, but would trigger the electron to release the stored-up energy as a second photon—as if the bucket was knocked off the door by the stream of water from a hose, resulting in a doubled downpour of water.

Nikolay Basov and Alexander Prokhorov found that photons of light of the right energy, in the nonvisible microwave region, triggered the release of further photons from ammonia. Generated in a sealed chamber, those photons could themselves stimulate yet more photons, a pyramid selling approach to producing light, not unlike a nuclear chain reaction. The result was something quite different from a conventional source of light. Because of the way they were stimulated, the light waves moved together, synchronized in their phase. It was the mechanism behind the device, amplifying the initial weak source of microwave photons, that led to its being described as microwave amplification by the stimulated emission of radiation—a "maser" for short.

By 1960, the American Theodore Harold Maiman had developed an equivalent device that worked with visible light. The concept had been the subject of a patent battle between American physicist Arthur Leonard Schawlow and another American, Gordon Gould. Gould was eventually recognized as the theoretical originator of the visible maser that Maiman was to build. Gould called his concept a laser, replacing the "microwave" in maser with "light."

Unlike the ammonia in the maser, Maiman's device contained a solid substance to produce the stimulated emission, a ruby, giving out a deep red light. The light was stimulated using a flash tube like a huge photographic flash unit. Inside the ruby, the light passed backward and forward, hitting mirrors at either end, each time stimulating more photons as the beam flashed back and forth. One mirror was only partly silvered, allowing part of the beam to escape while some remained in the system.

Because of the way that laser light is produced it is entirely different from the rays of the Sun or a lightbulb. The laser is a very powerful beam of light of a single color that is not easily scattered and dispersed as ordinary light is. A laser beam can be bounced off the Moon and will still return as a tight ray. And a beam of sufficient power can cut metal, bring down aircraft, and kill humans in good James Bond fashion.

While it's true that a laser has the potential to be a death ray, the key phrase there is "sufficient power." It takes a lot of energy to push out a laser that will do significant physical damage, so much so that we are unlikely ever to see vast lasers vaporizing whole cities. But with other unknown weapons they remain part of that classic sci-fi threat, the alien invasion. Is there any possibility this could become a reality? Could we be wiped out by UFOs?

For this to happen, we need to find some aliens—and that is a search that has proved surprisingly hard. It's said that Enrico Fermi, whom we last met developing the first nuclear reactor in Chicago, was once seated in the canteen at Los Alamos in the 1950s with three other physicists. Talk turned to UFOs, or "flying saucers" as they had just become popularly known. Fermi suddenly said, after some thought, "Where is everybody?"

He was reflecting on the lack of alien visitors. It might seem from all that has been said and written about UFOs and alien abductions that this was a silly question, as the truth (as the *X-Files* has it) is out there. Yet all the evidence is that the vast bulk of reported alien visitations have been misunderstandings or pure fantasy. It's telling that when the term "flying saucer" was first used, it was meant to describe how the craft moved (the way a saucer moves when it's skipped across water), rather than the shape of the ship. All the sightings since of saucer-shaped spacecraft seem to have been imaginings based on this misunderstanding.

The reason scientists like Fermi are surprised by the lack of alien visitors landing on the White House lawn is not because of all the dubious UFO sightings, but rather because of the sheer number of stars in the universe. It seems very likely that a universe on the scale we know ours to be should have many planets supporting life, some of which, we would expect, would have civilizations more technologically advanced than ours. Theories for the lack of contact break down into three broad categories: the aliens aren't there, they haven't found us, or they have found us but choose not to be seen.

Each of these ideas has its appeal. It is certainly true that the circumstances for life on Earth are quite specific, and it may just be that there are very few planets where life has developed be-yond the scale of bacteria—perhaps just one in our galaxy. (The scale of space is such that it's quite possible for other galaxies to be teeming with life without anyone ever reaching us in the Milky Way.) But scientists are wary of anything that suggests a special place for the Earth, when there is no good reason for it to have that special treatment. Of course, if there were only one planet with life, then it would be bound, by definition, to be the planet on which the inhabitants were thinking "Why us?"—but it still seems unlikely.

The size of the universe suggests the second possibility. It could be that there is plenty of life out there, but there is so much space that with the fundamental physical limit of not being able to travel faster than light, the aliens just haven't arrived yet. After all, there are plenty of parts of the oceans on the Earth that we haven't got around to visiting, and by comparison with space, the oceans are tiny. Perhaps the majority of space will never be visited by living beings. The only problem with this idea is that in such a scenario, there could still be some alien races that had managed to build self-replicating probes.

These devices, rather like the nanobots in chapter 6, would be capable of duplicating themselves from the raw materials they found around them. A probe would travel to a planet, duplicate itself perhaps many times, and then the new probes would fly off to more planets, spreading through a galaxy below the speed of light, but still using the power of doubling to quickly take in more and more of the galaxy. However, the probes would have had to start out many thousands of years ago to have spread far across the galaxy, and it's entirely possible that if one did visit the Earth, it wouldn't arrive in the tiny window in the Earth's lifetime when human beings would have been around to notice it.

The final possibility is that aliens know we're here but don't want us to be aware of them. Perhaps they have cloaking technology to be able to move among us unseen. Perhaps they have some kind of noninterference directive, or simply regard us as too unpleasant or inferior to mix with.

If either the second or the third case holds true, we could still at some point encounter aliens—and it's possible that like the aliens of so many B movies, they will prove unfriendly and will want to exterminate us. Equally, they could have friendly intentions. But in either case, given the lack of evidence to date, I'm not holding my breath waiting for alien invaders.

Those imagined probes could themselves prove dangerous to the human race, if they carried some kind of interstellar plague, or destroyed life on Earth in the process of replicating themselves— and they aren't the only kind of killer machines that science fiction has dreamed up.

Over the years we have increasingly put our day-to-day lives in the hands of machines, and a number of writers have considered what might happen if the machines decided it was time for them to run things. The movie series *Terminator* is based on this premise, one that has a noble lineage in written science fiction. In some stories it's the computers that take over. If computers reach the stage that they are truly thinking devices and consider themselves superior to us, will they leave us here, or throw us away?

This is a rather different picture from the Singularity, where machines and men come together to form a new species. The "rule of the machines" idea is rather that our everyday devices, from cars to air-conditioning plants, are becoming more and more intelligent as we incorporate more computing power in them. All the machines around us could gradually come to regard the human race as an inconvenience, or as something that is best looked after by keeping it docile—turning human beings into something close to pets.

I don't think we have a lot to worry about here yet. Just as was the case with robots and cyborgs, self-aware computers are probably further away than most technology enthusiasts predict, and we are generally quite good at building machines with sufficient built-in safety that they are unlikely to cause mass destruction, even if a few were to become rogues.

Other science-fiction disasters of the future have concentrated on the loss of vital resources. Water is one case of this. As we've seen in the chapter on climate change, we could face terrible water shortages in the future, but one science-fiction writer came up with a very different water-based threat to our survival. He imagined that we might lose our liquid water altogether. Liquid water is an essential for life, and the fact that it exists at all at the temperatures we need for life is due to a strange behavior of the water molecule. Water molecules are attracted to one another like little magnets, with the positively charged hydrogen being attracted to the negative oxygen in a different molecule. (This kind of attraction is called a hydrogen bond.)

The effect of this bond is that water molecules stick together more than you might expect. This makes water boil at a higher temperature than it would otherwise. Much higher. Water boils at 100 degrees Celsius (212 degrees Fahrenheit) at sea level. (The boiling point falls as air pressure drops, and it rises with higher pressure, which is how pressure cookers work. The increased pressure in a pressure cooker means the cooking takes place above 100 degrees Celsius.) If it weren't for hydrogen bonding, the boiling point of water would be well below −70 degrees Celsius (−90 degrees Fahrenheit). Water just wouldn't exist as a liquid on the Earth—and no water means no life.

The means Kurt Vonnegut devised to take away our water was to take something like hydrogen bonding even further. He imagined a special form of ice called Ice Nine that was so stable that it melted at only 45 degrees Celsius (114 degrees Fahrenheit). For most of the planet, water would be a permanent solid. Should a seed crystal of Ice Nine be dropped into a lake or an ocean it would spread uncontrollably from shore to shore, locking up the water supply and devastating the Earth.

Luckily, Ice Nine doesn't exist (though it is a wonderful concept), although there is a type of ice that forms at very low temperatures with the intentionally similar name of Ice IX. This, however, isn't stable at room temperatures, and presents no danger to our water supply or our survival.

However, ice has certainly put species at risk in the past. At one time, the Earth was largely tropical, but prior to the current spate of global warming, the general trend in temperatures on the planet had been downward for millions of years. The main influence for the underlying trend can be traced back to a geological event over 50 million years ago, when the tectonic plate supporting India ground its way into the Asian plate.

The result was a gradual change in the landscape as the Himalayan mountain range and the Tibetan plateau were brought into existence, thrusting higher and higher until they towered more than two miles above sea level. This is a big geological structure, covering the same area as about half of the United States. It was to have a major impact on the climate. The new structure interfered with the jet stream, the fast-flowing bands of air that circle the globe in the upper atmosphere.

Part of the impact was to change rainfall patterns, increasing the tendency to monsoon rains in the area. And, remarkably, the uplift of the Tibetan plateau also resulted in a kind of reverse of our current problems with climate change. Reactions of the rainfall on the rocks caused carbon dioxide to be taken out of the atmosphere. This reduced the greenhouse effect, lowering temperatures.

The overall trend of cooling that came from this vast geological event pushed the natural temperature cycles the world goes through below a freezing threshold. Over time, the Earth undergoes cyclical changes in its orbit, its rotation, and the degree to which it tilts. Bearing in mind just how much difference there is between winter and summer, brought on solely by the tilt of the Earth, it is hardly surprising that with the push downward in temperature caused by the geological changes, the temperature cycles could produce a terrifying result: the ice ages.

This isn't a new phenomenon. There have been many ice ages when ice sheets covered a major proportion of the Earth, but the most recent one, which we are technically still in, lasting around 2.5 million years, is the one that we know most about. During the worst of these so-called glacial periods, of which there have been around eighty during this ice age, driven by those tilts and wobbles in the Earth's path, sheets of ice have advanced to cover North America and Europe, making much of our present-day world uninhabitable.

Even when the ice didn't extend that far, during the glacial events, much of the life in the temperate zones was disrupted. Ice sheets blocked rivers, stopping them from flowing into the sea, causing massive flooding of coastal regions. What had been forests became icy tundras where few of the existing species could survive. And where there was new ice and snow, it encouraged further cooling. Just as positive feedback encourages climate change now, the positive feedback from the ice sheets had an effect. The more ice there was, the more sunlight was reflected back without warming the surface. This made it cooler, enabling more ice to form.

The first clues to this very different climate came from unexpected boulders. In the eighteenth century it was noticed that there were boulders in Alpine valleys that were far away from the native rock that had spawned them. It was known that glaciers carried rocks like these, and the only explanation seemed to be that the glaciers had once extended much farther than they now did. As geological science improved, there were other, more subtle signs of ice sheets traveling much farther than they now did, from linear scratching on rocks to the chemical analysis of different geological layers and the distribution of fossils. All indicated a regular visitation and withdrawal of the ice sheets over thousands of years.

In the interglacial periods, such as the one we now occupy, the ice sheets withdraw, leaving only residual sheets on land areas like Antarctica and Greenland. These interglacial periods have often lasted around ten thousand to fifteen thousand years—about how long we have been in this interglacial period; so there was some concern until recently that we might soon be plunged back into a glacial period, with the ice rendering the United States, Europe, and much of Asia uninhabitable.

In fact, this fear rested on an oversimplification; all the evidence was that even before the impact of climate change, we were probably several thousand years away from any danger from the ice. The problems of climate change mean that this is even less of a threat. Nonetheless, we should be aware that sometime in the future temperatures will begin to drop again, and ice will once more threaten human existence—though exactly when that will occur is not clear.

With all the possible threats we face from the natural world, from our science and technology, from war and terrorism, it might seem that we should be looking to the future with fear. On a bad day, it appears almost inevitable that we are going to wipe ourselves out. Yet there is still hope for the human race. We have to remember that science has been anything but all bad for us. And, as we'll see, the Pandora's box we opened has brought good as well as evil.

CHAPTER TEN
CAUTIOUS OPTIMISM

||||||||||||||||||||||||||||||||||||||||||||||||||||||

> *It is science alone that can solve the problems of hunger and*
> *poverty, of insanitation and illiteracy, of superstition and*
> *deadening custom and tradition. . . . The future belongs to science*
> *and those who make friends with science.*
>
> Jawaharlal Nehru (1889–1964), quoted in
> *The Making of Optical Glass in India, Proceedings of the*
> *National Institute of Sciences of India* (1961)

When we look back over the 4.5 billion years or so that the Earth
has been around, the vast majority of life-threatening disasters
that have brought a form of Armageddon to the world have been
natural rather than man-made. This is hardly surprising when you
consider the mere pinprick in time that is Homo sapiens' portion
of the Earth's existence. Allowing a very generous million years
(most figures put the emergence of Homo sapiens between 100,000
and 200,000 years ago), we have been around for one 4,500th of
the life of the Earth. Humanity's significant civilizations have existed for about a millionth of the lifetime of the Earth.

This means that we have, so far, missed out on the megadisasters that have resulted in mass extinctions of species. Looking
back in time, using the fossil record as a time telescope, it is pos-

sible to deduce events, known as extinction events, when immense swaths of life were wiped out on the Earth. The best-known we have already considered—the K/T event, which took place around 65 million years ago. At the boundary between the Cretaceous and Tertiary periods, the bulk of the dinosaurs perished.

The dinosaurs weren't alone. Around 50 percent of genera (the classification above species in the hierarchy of the description of living things) disappeared. This wasn't a uniform extinction—or we wouldn't be here. The conditions that ended the dinosaurs were survived by enough varied species of mammals for us to finally come out of the mix.

There have been at least four other major events, all in the last 500 million years. This doesn't mean that this period has been any worse than the years before it, but rather shows the limitations of the fossil-record telescope. There will have been events much longer ago, but we can't get a picture from the fossil record, as fewer and fewer fossils are available, and for a considerable time the only life was microscopic. The oldest of the known events was the Cambrian/Ordivician, around 488 million years ago, followed by the bigger Ordovician/Silurian event around 450 million years ago, which wiped out around 57 percent of genera.

Then came the late Devonian event, around 370 million years ago, followed by the massive Permian/Triassic event, sometimes given the evocative name "the great dying." Occurring 250 million years ago, this massive shock to the Earth's living systems wiped out over 80 percent of genera, taking an even greater toll of marine species, where a remarkable 96 percent were wiped out. It's the only mass extinction that has had a significant effect on insects as well as other creatures. In many ways this was the inverse of the

K/T event, with many mammal-like reptiles wiped out, while dinosaurs were given the chance to come into the ascendant.

Finally, before the K/T, there was the Triassic/Jurassic event around 205 million years ago, with just under 48 percent of genera being eliminated. Small by comparison with some of the others, this was still a terrible toll when you consider that practically half of the families in existence were wiped out. These five mass extinctions were just the massively large-scale events, with other smaller groups of extinctions punctuating the periods in between.

If we consider the whole of human existence it has also typically been the natural events that have most significantly reduced the population, whether it was ice age or earthquake or plague. But in the last hundred years we have seen our ability to reap mass destruction from our science and technology blossom like a horrible flower.

The First World War killed around 10 million military personnel and more than 6 million civilians. The "Great War" was considered the war to end all wars because of the sheer horror of the mortality numbers, combined with the even greater casualty count. But just twenty-one years later, the Second World War dwarfed that level of slaughter. This would see more than 22 million military deaths and somewhere between 34 million and 47 million civilian deaths, a total death toll probably exceeding 60 million. This is comparable in scale to the Black Death, which is believed to have killed up to 50 million in Europe (though the population was much lower in the fourteenth century, so the percentage who died was much greater). With the two world wars, science and technology proved all too well their capability as a mass destroyer.

Outside of war, industrial accidents have provided our biggest man-made disasters. The Chernobyl disaster may have resulted in as many as four thousand deaths, though this is hard to verify, as much of the mortality ascribed to the reactor explosion took place many years later, when the initial cause was hard to prove. More certain is the direct link between the Bhopal disaster in India in 1984 and its casualties. This certainly killed four thousand and may have resulted in as many as twenty-five thousand deaths.

The American chemical giant Union Carbide (now subsumed into the Dow Chemical Company) ran a large chemical plant in Bohpal, a city in the Madhya Pradesh state of India. On the evening of December 3, 1984, storage tank number 610 at the site, containing forty-two tons of the dangerous chemical methyl isocyanate, used in the production of pesticides, was contaminated with a large quantity of water. It is still not certain how that water got into the tank.

In the ensuing reaction between the water and the methyl isocyanate, temperatures in the tank soared to well over the boiling point, reaching around 200 degrees Celsius (400 degrees Fahrenheit). The temperature and ensuing pressure far exceeded the tank's capability to contain them. To avoid an explosion, the tank automatically vented gas, sending huge quantities of poisonous fumes into the atmosphere. Around half a million people were close enough to the plant to be affected by the gases.

Thousands died in their beds. Many more were injured in the struggle to get away from the area, or by inhaling the fumes. As well as methyl isocyanate, the population was exposed to a range of other noxious gases, from nitrous oxide to phosgene, as the overheated chemicals furiously reacted with the atmosphere. Union

Carbide has since paid out millions of dollars in compensation, but maintains that it was not responsible for the accident, blaming sabotage by disgruntled workers. Whatever the initial trigger really was, it was the location of the plant that has to be seen as the most significant factor in this tragedy.

Not all the deaths that have arisen from our use of science and technology have been on a large scale, but they should not be ignored. There have also been small-scale tragedies like the death of Marie Curie already mentioned, and near misses like the partial reactor core meltdown at Three Mile Island in Pennsylvania. These might not be worthy of the "Armageddon" tag in their own right, but they indicate circumstances that could have resulted in much wider danger. Scientists don't always have a great track record in keeping themselves and others safe.

There is no doubt that science, particularly the application of science through technology, carries with it dangers for humanity, dangers that time and again have come to terrible fruition. The old myth of Pandora's box was never truer than it is now—in fact, if it hadn't been dreamed up all those years ago, it would be necessary to create it today. The term "Pandora's box" has become a cliché. So it's worth spending a moment on just what the original story was before seeing how applicable it really is.

In ancient Greek mythology, Pandora was the equivalent of Eve, the first woman, created directly by the gods. She was given by her creators a jar that was never to be opened, containing all the ills of the world, from disease to suffering. This jar would become the legendary box due to some careless mistranslation in the sixteenth century. The jar was, in Greek, called a *pithos*, from which we probably get the word "pitcher."

The ancient Greek poet Hesiod had written an epic poem called *Works and Days,* which retold many of the Greek creation myths. In it, he referred to Pandora's *pithos.* But when the poem was translated to Latin by the medieval Dutch scholar Erasmus, he mistakenly turned *pithos* into *pyxis,* which is Latin for "box." So the original Pandora myth was closer to the Arabic stories where evil djinns are locked away in jars.

But whatever the container, the intensely curious Pandora opened it. She had to learn what was inside. As the stopper was removed, all the evils that beset us flew out to infect mankind, leaving behind in the jar only hope. (I'm not clear why hope was in there in the first place, but its symbolic role in Pandora's actions is clear.)

It's easy to see parallels between Pandora's opening of the jar and the story of the apple in Genesis. Here, in the second of the creation myths presented in the Bible, again it is the first woman who lets loose something dangerous, though in Eve's case it is a more subtle danger. I need to be clear about what's meant by myth here, as the word is often used in a derogatory way, and that isn't my intention. A myth is a narrative with a purpose. It tells of something useful for our everyday life through a story, usually occurring far in the past or in a distant land. The originator of the myth uses this exotic setting to explain a universal truth, or to put across an important piece of information in a way that will make it easier to remember and absorb.

In chapter 2 of Genesis we hear that there was one special tree in Eden. God tells Adam, "You may eat from every tree in the garden, but not from the tree of the knowledge of good and evil; for on the day that you eat from it, you will certainly die." Later, the newly created woman is told by the serpent, "Of course you will not die. God knows that as soon as you eat it, your eyes will be opened and you will be like gods knowing both good and evil."

Eve likes the look of the fruit (never named as an apple in the Bible) and eats some of it, also giving a portion to Adam—so they are cast out, though strangely, the serpent's version of events seems closer to the truth than God's, because they don't die on the day they eat the fruit.

In both these myths, curiosity, or the desire for something unattainable, perhaps something for which the human being is unworthy, produces suffering for humanity. Our innate desire to know more, to investigate and to learn, rewards us with pain. And there is no going back. We can't close the box (or rather, reseal the jar), nor can we unbite the apple. There is no way to return to an age of innocence. As the second law of thermodynamics has it, entropy increases.

Yet surely, the mythical attitude to discovery reflects a misunderstanding that enables us to end this book on a note of cautious optimism. It's true that curiosity in a physically dangerous environment leads to risk. Yet it also leads to great reward. There was no Garden of Eden in the real world. We don't have an idyllic past, which we could return to if we only abandoned all the advances of science and technology and became once more scavengers on nature's bounty.

The fact is that most of the time, nature is pretty harsh on those who suffer it without any mitigation. Sustenance is rarely plentiful, and there are many physical dangers that face us. Science and technology have mostly proved either harmless or vastly beneficial to humanity. The world has gained hugely from scientific discoveries in everything from medicine to information technology.

Yes, we have created terrible weapons of destruction, and we are responsible for the threats that arise from our abuse of the planet, bringing the very real menace of climate change. But we also live more comfortable lives, protected from many of the medical threats that are an everyday reality for those limited to a natural world. It has been pointed out that the life span of a healthy human hasn't altered vastly since biblical times, when three score years and ten (seventy) was given as the target age—but this hides a terrible reality.

You only have to think of the changes in child mortality over the years. Bear in mind that until well into the nineteenth century the majority of babies would die before reaching adulthood. When there's a funeral for a baby or a child it is always a very emotional and particularly sad occasion. It is sobering to think that not many years ago, and throughout all of history before that, the majority of funerals were for babies and children.

Perhaps even more significant than the medical benefits science has brought us are the changes in the quality of life that we experience. For practically every human being, until very recently in terms of human evolution, life was one long struggle. There was no time for enjoyment, for wonder, for all the things that arguably make life worth living. There was just the endless fight to keep food and drink coming in, to reproduce, and to avoid predation.

Sadly, for a shocking percentage of the world's population this is still the case. But for those of us lucky enough to live in the rich world, we have a transformed life, largely thanks to science and technology. Just read the quote from former Indian prime minister

Nehru at the opening of this chapter: "It is science alone that can solve the problems of hunger and poverty, of insanitation and illiteracy, of superstition and deadening custom and tradition. . . . The future belongs to science and those who make friends with science." This was no Western fat cat, selfishly appreciating what he had, but someone who was well aware of the need for the benefits that science could bring.

What's more, despite the caricature of emotionless beings driven by faulty logic, scientists are human too. In fact, there is something facile and inconsistent about the way that scientists are often portrayed as monomaniacs who ignore the consequences for the human race that inevitably follow from their evil work. On the one hand, scientists are portrayed as being cold, calculating, virtually inhuman in their logic. On the other hand, they can't follow through as simple a logical chain as "Produce something that kills lots of people and lots of people may well be killed."

There have been many examples of scientists contemplating the moral implications of their work, and often society exerts considerable controls to restrain scientific endeavor and to avoid unnecessary risk. Take the whole area of human genetic manipulation. In all countries where such work is carried out there are strict rules and regulations, with agonizingly careful consideration of each problem that is thrown up by any new developments. We might not all agree with the approach taken by another country's genetic ethics, and there will always be the occasional mavericks, like those who claim (almost certainly falsely) to have produced human clones, but the reality is that science does tend to be better policed than most areas of human endeavor.

It's arguable that the picture of scientists as emotionless, uncaring individuals reflects the strong correlation between being involved in science and being on the autistic spectrum. Once, autism was a label for the debilitating extreme. Thirty years ago, half the children diagnosed with autism couldn't speak, and the majority were below average intelligence. Those with such devastating autism find it practically impossible to communicate, living in an isolated world of their own. However, by the 1990s it was realized that these individuals were the tip of an iceberg, and that many more people were at the high-functioning end of the spectrum.

Autism is largely a genetic condition—if one identical twin has autism, there is a better than 50 percent chance the other will too. Being on the autistic spectrum is also neurodevelopmental: some modules of the brain are formed differently during a baby's early development. It is a real, physical condition. But we need to get away from the stereotype of low-functioning individuals or idiot savants like the character portrayed by Dustin Hoffman in the movie *Rain Man*. They exist, but they are not the norm. Here is Hans Asperger, for whom a form of autism, Asperger's syndrome, is named, commenting on the impact of autism:

> We have seen that autistic individuals, as long as they are intellectually intact, can almost always achieve professional success. . . . A good professional attitude involves single-mindedness as well as a decision to give up a large number of other interests. . . . It seems that for success in science or art, a dash of autism is essential.

According to brain expert Simon Baron-Cohen, "The female brain is predominantly hard-wired for empathy. The male brain is predominantly hard-wired for understanding and building systems." Autism, which Baron-Cohen describes as having an extreme male brain, brings with it an urge to collect information, classify, and systematize that drives many on the high-functioning end of the spectrum into the sciences. Despite all the encouragement to take up math and sciences now given to girls, the ratio of males to females in the physical sciences is still nine to one. (It's an interesting comparison that the ratio of males to females with high-functioning autism or Asperger's syndrome is ten to one.)

I am not suggesting that all scientists are on the autistic spectrum, or that having a mild case of the condition is a requirement to be an effective scientist. Yet it does seem that enough scientists have a toe in the water with a mild, high-functioning type of autism that one symptom of the condition could be responsible for the uncaring image that scientists have. As well as the single-mindedness Asperger mentions, autism is strongly associated with a lack of empathy.

This doesn't have to be as bad as it sounds. It's unfair to say that those with an extreme male brain are incapable of relationships; certainly those at the high-functioning end of the spectrum are capable. Such people can get a lot out of being with others. They are often embarrassingly loyal defenders of individuals and causes they support. Nor are they coldhearted, without a concern for others. The problem with the extreme male brain is its inability to read emotions and to empathize with another individual.

The result is often shock and distress on the part of the person on the autistic spectrum when hurt has been caused, in part because he can't understand why his actions have upset anyone. "It's not my fault!" is the frequent refrain in a relationship with such a person, genuinely bewildered as to why things have gone wrong and distressed at the problems he has caused.

When stereotyping the scientist, it is easy to attribute such a lack of empathy to coldness. We see an individual who is obsessed with rationality while putting other human beings at risk, just as Dr. McCoy is always needling the Vulcan Mr. Spock in *Star Trek* for his lack of empathy. But portraying scientists as cold is a misunderstanding of the condition. Whether or not on the autistic spectrum, scientists are no colder than anyone else, and they care as much about the implications of their work.

However much scientist care, we can never be absolutely certain that science won't end the world, nor can we be sure that it won't cause so much damage that human life in the future becomes much worse. Many aspects of science and technology now have the potential to cause devastation. We need to be grown up in the way we manage and respond to science. At the moment, the vast majority of people making significant decisions about our world do not have a good education in science. Until now it has been seen as good not to know anything about the scientific world. It has somehow been considered better, more intellectual, to focus on the arts.

To quote the British historian Lisa Jardine:

> Fifty years on, [C. P.] Snow's ominous prophecy of a governing class lacking the competence to make informed policy choices where science and technology are concerned continue to reverberate. In recent debates about GM crops, nuclear energy and climate change, politicians and the public have shown themselves liable to be swayed

> by the most persuasive of advisors or interest groups,
> unable to judge for themselves either the soundness of the
> scientific data or the scientific arguments.

We can no longer afford to take the stance that an educated person doesn't need to know about science. Science is largely responsible for keeping us alive, and has the potential for mass destruction. Every person who is making decisions in the modern world should ensure that he or she knows more about science. People need to have access to the right information to make sensible judgments about topics where science has a role to play—and that has become pretty well all of life. People need an education that includes a good understanding of science in all disciplines.

We also have to accept that there will always be rogue individuals. There will be people who lack the capability to make moral decisions, who consider their own fame and success more important than scientific values, or who are so obsessed with a particular technology or viewpoint that they will attempt to misuse technology, to throw science back in the face of civilization. This isn't a reason to walk away from science. Just because fire can cause terrible damage is no reason not to have heat when it's cold, or not to use it to cook our food.

On the whole, the scientific establishment is quite good at weeding out rogues in its midst, even though the process can be painfully slow. There is greater danger where such mavericks become well established with those in political power, so that the decisions are being made by those with little knowledge of science. The classic case was that of Trofim Denisovich Lysenko. Lysenko was a Russian biologist who rejected genetics in favor of the concept of acquired characteristics.

This was an idea advanced in the eighteenth century by French scientist Jean-Baptiste Lamarck. His thinking was not unreasonable, though it has been proved false. He assumed that animals and plants acquired characteristics from the stresses and strains they were placed under during their life, and that these characteristics could then be passed on to future generations. So, for instance, it seemed to him reasonable to assume that the ancestors of the giraffe would have stretched and stretched to try to get to high-up, delicate leaves. The result, Lamarck thought, would be a gradual lengthening of the animal's neck as a result of this repeated stretching. Subsequently, he thought, its children would tend to have rather longer necks than usual because the parents had, and so, gradually, the familiar giraffe neck would form.

Contrast this with the Darwinian idea that those giraffes that naturally had longer necks would tend to have a better chance of surviving, passing the tendency to have longer necks on to their offspring. In the conventional evolutionary approach, the animals that survive to breed are the ones that have randomly grown up with a longer neck. In the Lamarckian view, the necks actively stretched during a single animal's lifetime.

Lamarck had merely devised a theory, but in Stalin's USSR, Lysenko was able to impose his ideas on the agricultural community, resulting in disastrous decisions that had a direct impact on the food-growing capability of the Soviet Union. For example, Lysenko believed that as a result of acquired characteristics, it was possible to improve wheat yields vastly by putting the plants into stressful conditions; in fact, the result was just poor crops.

Lysenko's iron hand on the scientific establishment meant that right through to the 1970s, Russian biological sciences were well behind their Western counterparts. They failed to pick up on the advances in genetics that were transforming biology elsewhere, seeing this as false, capitalist science. If Lysenko had simply been a scientist with wacky ideas, his theories would have been tested against reality, the scientific establishment would have sidelined him, and the genetic explanation would have triumphed much sooner—but this rogue scientist's links to those in political power overrode any scientific sensibilities.

Some would also say that Edward Teller's role as champion of the hydrogen bomb was another example of a rogue scientist who made dangerous advances because of his influence in political circles, rather than for any great scientific value in his theories. It's certainly true that many other scientists were deeply unhappy about the deployment of thermonuclear weapons, seeing them as a step too far, an argument that Teller never really had to refute because of his connections with those in political power. (Teller would later play a similar role in the development of Ronald Reagan's "Star Wars" Strategic Defense Initiative for a missile defense shield.)

Some scientists might be clearly identifiable as taking a wrong tack because they are inherently "bad." Others are guilty of nothing more than being human. Human beings make mistakes. There will be errors of judgment and accidents. The possibility of accidents occurring is generally not a good argument to avoid taking action—otherwise we would all sit at home under the table all day, as there is risk in stepping out of the house. However, those whose actions can have an impact out of all proportion to their personal scale do need extraordinary checks on their behavior.

We can never totally eliminate risk, but we can put procedures and controls in place to minimize that risk. As accidents at nuclear plants have shown, there is always a human tendency to assume things will work out okay, and so to circumvent safety systems that are tedious to operate. What's more, there is plenty of evidence that big business will try to reduce the impact of safety systems on their profits by keeping safety efforts to a minimum, and avoiding official monitoring (inevitably described as government bureaucracy) wherever possible.

There is no doubt that we need to keep up the pressure on those responsible for potentially deadly science and technology to ensure that all sensible safety measures are taken. It will never be a matter of 100 percent certainty. We can't be sure all life on Earth won't be wiped out by a meteor collision or a gamma ray burst. But we can keep a handle on the dangers we do have the ability to influence.

Once again, what's important is that we improve the understanding of science and scientists among the population as a whole, so that the voting public can make the decisions that control science wisely, rather than be inspired by bogeymen and half-truths. Just to show how it's possible for decisions to be made based on rumor and fear rather than a sensible understanding of the facts, take the example of the scare that arose around the MMR (measles, mumps, and rubella) vaccination a few years ago. This was a classic case of the sort of reaction to science I call a bogeyman. This is a reaction to something that doesn't really exist.

When a bogeyman comes along, all balance goes out of the window. Perhaps the strongest bogeyman card anyone can play is danger to our children. When children are put at risk, our sense of balance and fairness is abandoned, thrown away in a moment in response to the natural parental concern for our offspring. Sadly, the media quite often raise the public's awareness of a bogeyman with insufficient evidence—and the inevitable response comes because we don't wait for detailed confirmation in dealing with a bogeyman, but go straight into panic mode. It's a scientific version of the lynch mob.

The MMR scare was started in 1998 by British doctor Andrew Wakefield. For nearly ten years following the release of his research, the suggestion that the measles, mumps, and rubella vaccine could cause autism in children frightened and confused the public. Millions of children missed out on vaccinations and were put at risk. Yet all this fuss was based on a flawed study of twelve individuals made by one semi-amateur.

Rather than listen to the wide range of experts who had undertaken vastly larger, more conclusive studies demonstrating the safety of MMR, the public was sparked into fear by the way the media picked up on the scaremongering of one man, producing stories based on little more than anecdote. And once a bogeyman has been raised, it is very difficult to keep it down. Ten years after the original publicity there were still occasional bursts of MMR panic in the media, despite outbreaks of measles among those not inoculated, causing serious illness and at least one death.

Part of the problem is that those who work in the news media, unlike scientists, are unwilling to reveal their mistakes. Science often advances by learning from error. But when contrary results show that a scare story was based on false evidence, as happened with the MMR panic, this is not reported with the same depth as the original story—if it is covered at all. What typically happens is that someone announces some research before a peer-reviewed paper comes out. The media pick up on this and make a big splash, terrifying everyone. Then either the paper isn't published, or later work shows that the original tentative findings were wrong. And the media respond with silence. We don't get to hear of the new research. We certainly don't hear "Sorry, we scared you unnecessarily; we got it wrong." The reality is just ignored.

The general public needs to better understand not only scientific theories, but the basic principles of science, including that beloved one of the bogeyman hunter, "'Data' is not the plural of 'anecdote.'" Just because we've heard something in a bar, or have heard the subjective experience of one individual, does not mean that we can make any useful initial judgments. There *is* smoke without fire. Arguably, anyone making important decisions about science and technology that can influence all our lives should be able to demonstrate that she is not susceptible to bogeymen, and that she understands the basics of science well enough to be able to make an informed decision.

In the end we have to face the fact that science has dangers. Of all the forms of knowledge, it has the greatest potential for harm and the greatest for good. It does make Armageddon at our own hands possible. But science brings us hope, too. More than that, it brings us insight. It responds to our curiosity. And curiosity is a natural and desirable trait. Despite the old saying about what curiosity did to the cat, most cats survive. And unlike cats, we can combine curiosity with wisdom and knowledge.

Treated correctly, the Pandora's box of science brings us much more than hope. We wouldn't expect to take a coal out of a red-hot fire with our bare hands so we could look at it or use it. In the same way, we need caution and the right protection when exploring the nature of the universe and the applications of the scientific discoveries we make. But to say that Pandora's box should remain shut is not the answer. To do so would be to miss out on many treasures.

Take care. But do open the box.

NOTES

CHAPTER I
MAD SCIENTISTS
||||||||||||||||||||||||||||||||||||||||||||

PAGE 1—The ability to imagine the future and its impact on human beings is described in Brian Clegg, *Upgrade Me* (New York: St. Martin's Press, 2008).

PAGE 2—The archbishop of Canterbury's sermon, making the first use of the term "weapons of mass destruction," is quoted in "Archbishop's Appeal," the *Times* (London), December 28, 1937.

PAGE 4—Mary Shelley, *Frankenstein: Or, The Modern Prometheus* (London: Penguin, 2004).

PAGE 6—Information on the suspicions about medieval philosophers dealing with the devil, and about talking metal heads, is from Brian Clegg, *The First Scientist* (London: Constable & Robinson, 2003).

CHAPTER 2

BIG BANGS AND BLACK HOLES

||||||||||||||||||||||||||||||||||||||||||||||||||||||||||||||||||||||||||||

PAGE 11—For more on the big bang and alternative theories see Brian Clegg, *Before the Big Bang* (New York: St. Martin's Press, 2009).

PAGE 12—The analogy between finding a watch on a heath and finding the complexity of nature is from William Paley, *Natural Theology* (Boston: Gould & Lincoln, 1860), available at Google Books.

PAGE 16—The suggestion that the U.S. Air Force is working on antimatter weapons and information on antimatter from Frank Close, *Antimatter* (Oxford: Oxford University Press, 2009).

CHAPTER 3

ATOMIC DEVASTATION

||||||||||||||||||||||||||||||||||||||||||||||||||||||||||||||||||||

PAGE 40—Sue Guiney's reminiscence of air-raid drills is from her blog, www.sueguineyblog.blogspot.com.

PAGE 41—Rutherford's fifteen-inch shell and tissue paper comparison is from www_outreach.phys.com.au.uk/camphy/nucleus/nucleus5_1.htm.

PAGE 41—Rutherford's remark that atomic power was moonshine is from the *New York Herald Tribune*, September 12, 1933.

PAGE 42—Leo Szilard's conceiving the nuclear chain reaction while crossing the road is described in his collected works, noted in P. D. Smith, *Doomsday Men* (London: Penguin Books, 2007).

PAGE 43—Information on the origins of nuclear fission and the development of atomic weapons during the Second World War is from Jim Baggott, *Atomic* (London: Icon Books, 2009).

PAGE 45—H. G. Wells introduced the term "atomic bomb" in *The World Set Free* (first published 1913; First World Library, 2007).

PAGE 46—Letter from Paul Harteck to the German war office on April 24, 1939, is described in Jim Baggott, *Atomic* (London: Icon Books, 2009).

PAGE 49—Glenn Seaborg's admission that the plutonium he showed off was often green ink is referenced in P. D. Smith, *Doomsday Men* (London: Penguin Books, 2007).

PAGE 54—Section of *Holy Sonnets* 14 from John Donne, *Poems of John Donne*, vol 1, ed. E. K. Chambers (London: Lawrence & Bullen, 1896).

PAGE 54—Robert Oppenheimer's letter about the origins of the name Trinity for the atomic bomb test site is quoted in David Quammen, *Natural Acts* (New York: W. W. Norton, 2008).

PAGE 55—Otto Frisch's account of the Trinity bomb test is from Otto Frisch, *What Little I Remember* (Cambridge: Cambridge University Press, 1980).

PAGE 56—Isidor Rabi's account of the Trinity bomb test is from Isidor I. Rabi, *Science: The Center of Culture* (New York: World Publishing, 1970).

PAGE 57—Paul Tibbets's recollection of being on the flight deck of the *Enola Gay* is quoted in Frank Barnaby, *How to Build a Nuclear Bomb* (London: Granta, 2004).

PAGE 58—The White House press release on Hiroshima is available at www.atomicarchive.com/Docs/Hiroshima/PRHiroshima.shtml.

PAGE 60—Information on Marie Curie is from Denis Brian, *The Curies* (Hoboken, NJ: John Wiley, 2005).

PAGE 64—The Franck report is described in Jim Baggott, *Atomic* (London: Icon Books, 2009).

PAGE 67—The plan for a preemptive strike on the Soviet Union formulated soon after the first atomic bombs were dropped is described in Arthur Krock, *Memoirs* (New York: Funk & Wagnall's, 1968).

PAGE 69—The majority and minority annexes from the General Advisory Committee's October 30, 1949, report on hydrogen bombs are from www.atomicarchive.com/Docs/Hydrogen/GACReport .shtml.

PAGE 71—President Truman's address announcing the work on the hydrogen bomb is available at www.atomicarchive.com/Docs/ Hydrogen/HBomb.shtml.

PAGE 72—Edward Teller's response to the H-bomb test is described in Edward Teller, *Energy from Heaven and Earth* (San Francisco: W. H. Freeman, 1979).

PAGE 74—Edward Teller's likening the risk of fallout to being an ounce overweight is quoted in Peter Goodchild, *Edward Teller: The Real Dr. Strangelove* (London: Weidenfeld & Nicolson, 2004).

PAGE 76—The observation that the permanent members of the UN Security Council all have nuclear weapons is from Frank Barnaby, *How to Build a Nuclear Bomb* (London: Granta, 2004).

PAGE 77—The *University of Chicago Round Table* show on atomic weapons is described in P. D. Smith, *Doomsday Men* (London: Penguin Books, 2007).

PAGE 81—Among the support for the practicality of the cobalt bomb was James Arnold's assessment in the *Bulletin of the Atomic Scientists* 6 (October 1950): 290–92.

PAGE 81—Tom Lehrer's song "We Will All Go Together When We

Go" is in Tom Lehrer, *The Tom Lehrer Song Book* (New York: Crown, 1954).

PAGE 81—General MacArthur's plan to make an unpassable zone between Korea and China is described in "Gen MacArthur Complained of British Perfidy" the *Times* (London), April 9, 1964.

PAGE 83—The Russian Perimetr automated defense system is described in P. D. Smith, *Doomsday Men* (London: Penguin Books, 2007).

PAGE 85—Luis Alvarez's comments on the ease of making a bomb with enriched uranium are quoted in Frank Barnaby, *How to Build a Nuclear Bomb* (London: Granta, 2004).

PAGE 87—The report of the Committee on Medical Preparedness for a Terrorist Nuclear Event is *Assessing Medical Preparedness to Respond to a Terrorist Nuclear Event: Workshop Report* (Washington: National Academies Press, 2009).

PAGE 90—The activities of the Nuclear Emergency Support Team and the various attempts at nuclear extortion the United States has faced are described in Jeffrey T. Richelson, *Defusing Armageddon* (New York: W. W. Norton, 2009).

PAGE 93—The use of Geiger counters around Three Mile Island is described in Richard A. Muller, *Physics for Future Presidents* (New York: W. W. Norton, 2008).

PAGE 96—The accusation by the chairman of the U.S. Senate Intelligence Committee that the Soviets lied about the cessation of the chain reaction at Chernobyl is mentioned in Richard A. Muller, *Physics for Future Presidents* (New York: W. W. Norton, 2008).

PAGE 96—The impact of the Chernobyl accident on local flora and fauna is described in Mary Mycio, *Wormwood Forest: A Natural History of Chernobyl* (Washington: Henry Joseph Press, 2005).

PAGE 98—Information on the pebble-bed design of nuclear reactor is from Richard A. Muller, *Physics for Future Presidents* (New York: W. W. Norton, 2008).

PAGE 99—Information on nuclear fusion power generation from Brian Clegg, *Ecologic* (London: Eden Project Books, 2009).

CHAPTER 4

CLIMATE CATASTROPHE

|||||||||||||||||||||||||||||||||||||||||||||||||||||||||||||||||||||

PAGE 102—Richard Turco's decription of nuclear war causing climate change is from "Nuclear Winter: Global Consequences of Multiple Nuclear Explosions," *Science* 222 (1983): 1290.

PAGE 104—Senator Barbara Boxer's discovery of payments for articles, Tim Worth's comparison with the tobacco industry and the Exxon meeting are described in Sharon Begley, "The Truth About Denial," *Newsweek*, August 13, 2007.

PAGE 105—Information on climate change and its impact from Brian Clegg, *The Global Warming Survival Kit* (London: Transworld, 2007) and *Ecologic* (London: Eden Project Books, 2009).

PAGE 130—Details of the May 2009 U.S. scheme to reduce car emissions are from the BBC Web site, http://news.bbc.co.uk/1/hi/world/americas/8056908.stm.

PAGE 131—Zeolitic imidazolate frameworks to capture CO_2 are described in Andy Coghlan, "Crystal Sponges Capture Carbon Emissions," *New Scientist*, February 23, 2008.

PAGE 132—Methane being twenty-three times more powerful a greenhouse gas than carbon dioxide is described in Dave Reay, *Climate Change Begins at Home* (London: Macmillan, 2005).

PAGE 132—The assertion that livestock contributes 18 percent of greenhouse emissions is made in Bijal Trivedi, "How Kangaroo Burgers Could Save the Planet," *New Scientist*, December 25, 2008.

PAGE 133—Ken Caldeira's comments on solar shields are from Catherine Brahic, "Solar Shield Could Be a Quick Fix for Global Warming," *New Scientist*, June 5, 2007.

PAGE 135—The Russian proposal to release 600,000 tons of sulfur into the atmosphere is described in Catherine Brahic, "Earth's Plan B," *New Scientist*, February 28, 2009.

CHAPTER 5

EXTREME BIOHAZARD

PAGE 138—Information on the 2009 swine flu pandemic from the BBC News Web site, http://news.bbc.co.uk.

PAGE 140—The potential lethality of anthrax is taken from "an official American study" quoted in Frank Barnaby, *How to Build a Nuclear Bomb* (London: Granta, 2004).

PAGE 141—The German officer's likening the use of chemical weapons to killing rats is from Edmund Russell, *War and Nature: Fighting Humans and Insects with Chemicals from World War I to Silent Spring* (Cambridge: Cambridge University Press, 2001).

PAGE 142—The suicide of Clara Haber is described in P. D. Smith, *Doomsday Men* (London: Penguin Books, 2007).

PAGE 142—Details of the April 1915 gas attack are from P. D. Smith, *Doomsday Men* (London: Penguin Books, 2007).

PAGE 143—Winston Churchill's enthusiasm for the use of gas is described in Martin Gilbert, *Winston S. Churchill, 1917–22* (London: Heinemann, 1975).

PAGE 148—Secretary of State Shultz's observation in 1989 about the risk of terrorist use of chemical weapons is quoted in Frank Barnaby, *How to Build a Nuclear Bomb* (London: Granta, 2004).

PAGE 149—The story of the siege of Feodosia and the use of plague victims as a biological weapon is from Frank Barnaby, *How to Build a Nuclear Bomb* (London: Granta, 2004).

PAGE 151—Ken Alibek's assertion that the outbreak of tularemia among German soldiers attacking Russia was caused by an early biological weapon is made in Ken Alibek and Stephen Handelman, *Biohazard* (London: Arrow Books, 2000).

PAGE 155—Examples of terrorist groups using or intending to use biological weapons are from Frank Barnaby, *How to Build a Nuclear Bomb* (London: Granta, 2004).

PAGE 155—Methods for spreading biological weapons are discussed in Frank Barnaby, *How to Build a Nuclear Bomb* (London: Granta, 2004).

PAGE 156—Ken Alibek's comment that the manufacturing technique is the weapon is from Ken Alibek and Stephen Handelman, *Biohazard* (London: Arrow Books, 2000).

PAGE 158—Ken Alibek's assertion that the Soviets were developing a new biological weapon every year in the 1980s is from Ken Alibek and Stephen Handelman, *Biohazard* (London: Arrow Books, 2000).

CHAPTER 6

GRAY GOO

||||||||||||||||||||||||||||

PAGE 162—Michael Crichton, *Prey* (New York: HarperCollins, 2002).

PAGE 163—The early suggestions of the existence of atoms are from Brian Clegg, *The Instant Egghead Guide to Physics* (New York: St. Martin's Press / Scientific American, 2009).

PAGE 167—Richard Feynman's 1959 speech on nanotechnology, "There's Plenty of Room at the Bottom," is transcribed at www .zyvex.com/nanotech/feynman.html.

PAGE 169—For more on nanotechnology and assemblers see K. Eric Drexler, *Engines of Creation* (New York: Bantam, 1986).

PAGE 170—Bradley Edwards's comment on a space elevator is from an interview with Space.com at www.space.com/businesstechnol ogy/space_elevator_020327-1.html.

PAGE 171—The Soil Association's assumption that nanoparticles are safe if they are "natural" is reported in an editorial, "Natural Does Not Mean Harmless," *New Scientist,* January 26, 2008.

PAGE 172—The Soil Association's defense of its stance on nanoparticles is from an e-mail from Soil Association representative Gundula Azeez, dated August 4, 2008.

PAGE 181—For more on bees as a superorganism see Jürgen Tautz, *The Buzz About Bees* (Berlin: Springer, 2008).

PAGE 184—The NASA scientist predicting self-replicating robots by 2001 is quoted in K. Eric Drexler, *Engines of Creation* (New York: Bantam, 1986).

PAGE 186—Friends of the Earth's comment that having nanotech applications will confer huge advantages on the countries that control them is from the Friends of the Earth submission to the *Royal Society and Royal Academy of Engineering Study on Nanotechnology,* June 2003 (http://www.nanotec.org.uk/).

PAGE 186—Near-term military implications of nanotechnology are discussed in chapter 6 of the *Royal Society and Royal Academy*

of Engineering Study on Nanotechnology, June 2003 (http://www
.nanotec.org.uk/).

CHAPTER 7

INFORMATION MELTDOWN

|||||||||||||||||||||||||||||||||||||||||||||||||||||||||||||||||||||

PAGE 190—The rat and the workman disrupting communications
in New Zealand is described in "Rat Blamed for Latest Telecom
Blackout," *New Zealand Herald,* June 21, 2005.

PAGE 192—Information on ARPA and ARPANET is from Katie
Hafner and Matthew Lyon, *Where Wizards Stay Up Late* (New
York: Touchstone, 1996).

PAGE 193—Information on the ARPANET worm and Robert Morris is from Clifford Stoll, *The Cuckoo's Egg* (London: Pan Books, 1991).

PAGE 197—Information on the spread of the ARPANET worm is
from Charles Schmidt and Tom Darby, *The What, Why, and How
of the 1988 Internet Worm,* http://snowplow.org/tom/worm/worm
.html.

PAGE 198—Information on the etymology and first uses of "cybernetics" is from the *Oxford English Dictionary* (Oxford: Oxford
University Press, 1989).

PAGE 198—The small-scale examples of cyberterrorism are taken
from Dan Verton, *Black Ice: The Invisible Threat of Cyber-terrorism*
(Emeryville, California: McGraw-Hill, 2003).

PAGE 202—The research showing how failure of a subnetwork
with a low load could be better at taking out the U.S. West Coast
grid than failure of a subnetwork with a high load is at Jian-Wei
Wang and Li-Li Rong, *Cascade-Based Attack Vulnerability of the
US Power Grid,* Safety Science, 47, 1332, 2009.

PAGE 203—Ian Fells's comment about using Semtex near a power station is from Paul Marks, "How to Short-Circuit the American Power Grid," *New Scientist,* September 12, 2009.

PAGE 203—The 1997 cyberattack exercise is described in Dan Verton, *Black Ice: The Invisible Threat of Cyber-terrorism* (Emeryville, California: McGraw-Hill, 2003).

PAGE 203—The July 2009 cyberattack on U.S. computers is described on the BBC Web site at http://news.bbc.co.uk/1/hi/technology/8139821.stm.

PAGE 204—The number of serious cyberattacks on energy companies and the suggestion that many are organized from the Middle East features in Dan Verton, *Black Ice: The Invisible Threat of Cyber-terrorism* (Emeryville: McGraw-Hill, 2003).

PAGE 205—The threat from access to wireless networks, such as the American Airlines curbside check-in, is described in Dan Verton, *Black Ice: The Invisible Threat of Cyber-terrorism* (Emeryville, California: McGraw-Hill, 2003).

CHAPTER 8

NO LONGER HUMAN

||||||||||||||||||||||||||||||||||||||||||||||||||||

PAGE 209—The assertion that the urge to enhance ourselves is part of what makes us human is from Brian Clegg, *Upgrade Me* (New York: St. Martin's Press, 2008).

PAGE 211—Road traffic accident mortality rates from *Wolfram Alpha,* www.wolframalpha.com.

PAGE 212—Ray Kurzweil predicts the coming of the Singularity, merging human and technology to produce a new form of life, in Ray Kurzweil, *The Singularity Is Near* (London: Duckworth, 2005).

PAGE 213—The prediction of posthuman evolution via robots from BT's labs is from Ian Pearson, Chris Winter, and Peter Cochrane, *The Future Evolution of Man* (Ipswich, UK: BT Labs, 1995).

PAGE 214—Information on Neanderthal man is from Clive Finlayson, *The Humans Who Went Extinct* (Oxford: Oxford University Press, 2009).

PAGE 216—Damien Broderick's example of the growth curve of speed of transport is cited in Bill McKibben, *Enough: Staying Human in an Engineered Age* (New York: Henry Holt, 2003).

CHAPTER 9

FUTURE FEARS AND NATURAL PITFALLS

PAGE 222—John Michell's early ideas on earthquakes are described in Florin Diacu, *Megadisasters* (Oxford: Oxford University Press, 2009).

PAGE 225—The use of muons to probe the interior of Japanese volcanoes is described in "Cosmic Rays Reveal Volcano's Guts," *New Scientist,* October 3, 2009.

PAGE 226—Information on the Yellowstone supervolcano is from Bill Bryson, *A Short History of Nearly Everything* (New York: Broadway Books, 2003).

PAGE 233—Details of Archimedes' weapons are from Brian Clegg, *Light Years* (London: Macmillan, 2007).

PAGE 235—Bulwer-Lytton's vril is described in P. D. Smith, *Doomsday Men* (London: Penguin Books, 2007).

PAGE 236—For more on the Bulwer-Lytton Fiction Contest see www.bulwer-lytton.com/.

PAGE 238—Enrico Fermi's wondering where the aliens were is described in Marcus Chown, *We Need to Talk About Kelvin* (London: Faber & Faber, 2009).

PAGE 242—Ice Nine appears in Kurt Vonnegut, *Cat's Cradle* (London: Penguin, 1965).

PAGE 242—The impact of the rising of the Himalayas and the Tibetan plateau is described in Clive Finlayson, *The Humans Who Went Extinct* (Oxford: Oxford University Press, 2009).

PAGE 243—The most recent glacial period and the move into an interglacial one, with their impact on humanity, are described in Steven Mithen, *After the Ice* (London: Phoenix, 2004).

CHAPTER 10
CAUTIOUS OPTIMISM

PAGE 251—Eve's taking the fruit of the tree of the knowledge of good and evil is from the New English Bible (London: Oxford and Cambridge University Presses, 1970).

PAGE 255—Information on the autistic spectrum, the extreme male brain, and its frequency of occurrence in the sciences is from Simon Baron-Cohen, *The Essential Difference* (London: Penguin, 2004).

PAGE 257—Lisa Jardine's comments on the lack of competence of the ruling classes to make decisions about science and technology are from her C. P. Snow Lecture, delivered at Christ's College, Cambridge, England, on October 14, 2009.

INDEX